DON'T DRINK THE WATER

(without reading this book)

The Essential Guide to Our Contaminated Drinking Water and What You Can Do About It

by

LONO KAHUNA KUPUA A'O

D1291828

KALI PRESS

First published in 1996 by:
Kali Press
Box 2169
Pagosa Springs, CO 81147

Library of Congress Cataloging in Publication Data
Kahuna Kupua A'o, Lono
  Don't drink the water. The essential guide to our contaminated
  drinking water and what you can do about it.

  Catalog card # 95-82079
  ISBN 0-9628882-4-9

Edited by: Susan Tinkle
Design/illustrations by: Tom Grignon
Administration: Paige Mooy
Photos: David W. Harp (pp. iv, 6, 28); Lew Pence (p. 32)

ACKNOWLEDGEMENTS

I AM GRATEFUL TO all those incredible brothers and sisters who love me unconditionally and generously support me as I journey down this path.

I am particularly grateful to Lenny Blank for introducing me to the publisher, and who, along with my beautiful sister Debbie Gordillo, provided me the incredible computer system I use to write these works and bring them to my family.

Of course, I am particularly grateful to my beloved wife Ka Ike Lani for all her loving support. I have been deeply privileged to be able to spend the last 17 years of my life with her.

I am also humbly grateful to all of my brothers and sisters who refuse to give away responsibility for their lives to government and an elitist medical establishment. Huna teaches us that the person with the consequences is the person with the responsibility, and the person with the responsibility is the person with the authority. Although many engage in the illusion that they can give responsibility for their health and well being to others, there are those who know better, and continually work to affirm each person's inalienable right to choose their own path of healing.

To these courageous spirits I say *mahalo* (thank you) and I encourage you to continue in your sacred work. Please know that your *ohana* (spiritual family) is most grateful for your courage and effort.

— *Lono Kahuna Kupua A'o*

CONTENTS

"When the well's dry, we know the worth of water."

—Benjamin Franklin

INTRODUCTION

WATER IS OUR MOST PRECIOUS RESOURCE.
It is incredibly unique because its chemical structure
allows it to exist in liquid form at temperatures
where other substances of double its molecular
weight are gaseous. This special property is the
major reason water is called the universal solvent,
and the key to life.

Because water is such an excellent solvent, it
is like a chameleon, taking on the character of its
surroundings. This peculiarity makes water the
foundation of a rich biogeochemical cycle that is
responsible for the abundance and variety of life on
this planet. The geologic record makes it perfectly
clear that when the balance of this cycle changes,
the planet's life forms change with it.

Until recently, this cycle (called the Hydrological
Cycle) has remained in a delicate but stable balance.
For the most part, it has changed so slowly that life
forms have had plenty of time to adapt, thus
ensuring their survival. But the same system that
gives birth to and nurtures life has also given birth to

an exploding population of unique bipedal mammals with extraordinary intelligence. Known as humans, these creatures have managed to amass a level of technological expertise that eclipses anything previously known in the history of the planet.

Knowledge is power, and power, in the absence of wise direction, is dangerous. Human technological capabilities, developed in just the last 100 years, have disrupted delicate ecosystems that have existed in a finely-tuned balance for tens of thousands, if not millions, of years. When this balance is disrupted, life forms that depend upon and are a part of those ecosystems either die or find survival as a species in doubt. Humans are one of those life forms, but we have not, as yet, had time to understand the effects of the causes we have set in motion.

Our lack of concern for the long-term consequences of disrupting these delicate ecosystems promises to deliver great pain to our generation, as well as future generations that must inhabit this planet. We are all part of the one. What affects any part of our world affects us all.

Among too many people, short-term financial considerations outweigh the less obvious, but far more important, survival issues that revolve around ecological balance. The attitude of "I'll take mine now and let the future take care of itself" seems to be the unspoken, but practiced, norm. As

Grandfather Wallace Black Elk, a Lakota Sioux medicine man, says: "Didn't they make a mistake when they printed America's motto on its money? Shouldn't it say: 'In **Gold** We Trust'?"

Unwise choices by a technologically advanced, but unwise, human population have already resulted in the widespread destruction of many parts of the delicate ecosystem we call our home. In the last hundred years or so, the pollution and depletion of precious water resources, combined with the destruction of our rainforests, has resulted in the eradication of more species of plants and animals than in the previous 50,000 years. We may not want to think about it, but we are at the brink. We must learn to become responsible for the technological power we have created. If life as we know it is to survive on this planet, we must not allow financial gain and personal comfort to continue to dictate our choices. We must learn to respect and harmonize with our world and the delicately balanced ecosystems that comprise it, or we will be destroyed in our own web.

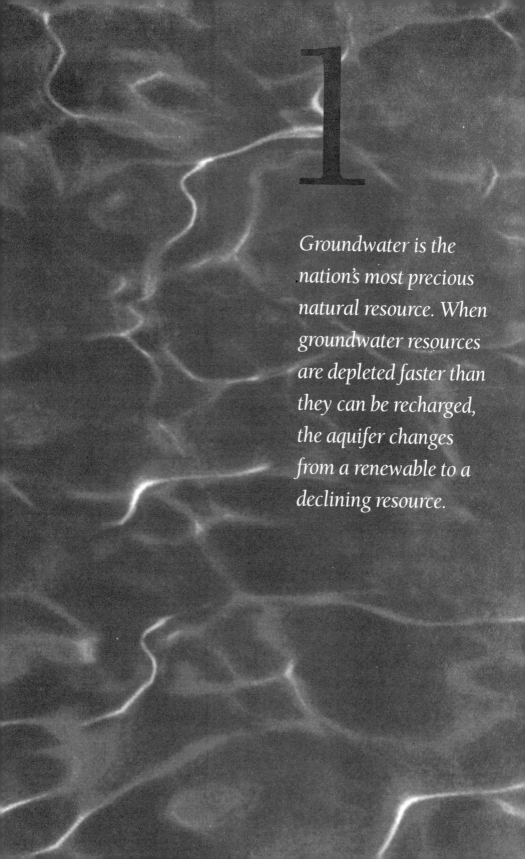

1

*Groundwater is the
nation's most precious
natural resource. When
groundwater resources
are depleted faster than
they can be recharged,
the aquifer changes
from a renewable to a
declining resource.*

WATER RESOURCES

IN ONE SENSE, THE EARTH'S water resources are inexhaustible, because water doesn't get "used up." The amount evaporated is exactly balanced by the amount that is precipitated. But water does change its character based on its experiences; the various parts of the earth's surface are involved in the cycle in very different ways and proportions, depending upon such factors as climate, geography, degree of forestation, etc.

Water is so abundant it covers over 70% of the earth's surface. It exists as a liquid in lakes, rivers, oceans and underground aquifers; it exists as a gas in the atmosphere; and it exists as a solid in the form of ice and snow. But if you were to compare all the water on earth to the amount of water in a gallon jug, fresh water available to sustain life would be just over one tablespoon, or less than half of 1%. About 2% of the world's freshwater supply is in the form of ice locked up in glacial formations. The balance, about 97%, is seawater.

THE HYDROLOGICAL CYCLE

The hydrological cycle refers to the natural circulation of water from the earth's surface into the air and back again to the surface. It is nature's way of cleaning the water, recycling it from the oceans to lakes, rivers, and streams on the surface, and recharging groundwater supplies. It is very effective, but very slow compared to the rate at which humans are polluting this precious resource.

The cycle begins with evaporation of surface water into the

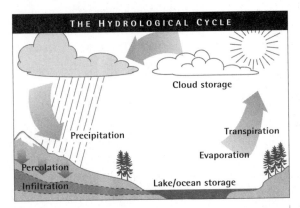

atmosphere. About 85% of all water evaporation comes from the oceans. The remaining 15% comes from lakes, rivers, and wetlands. This water is carried by the wind for tens to hundreds, or perhaps thousands, of miles. It then returns to the earth in the form of precipitation. Some of this water is used by plants in the process of photosynthesis and is quickly returned to the air. Some falls on the surface and runs off into the surface water systems of streams, rivers, and lakes. Some percolates through the soil to become groundwater, which moves very slowly back to the surface or to the oceans, completing the cycle. The majority returns directly to the oceans which cover 70% of the earth's surface.

THE UNIQUE NATURE OF WATER

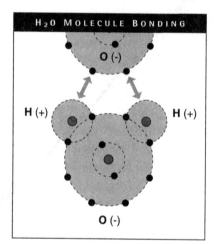

H₂0 MOLECULE BONDING

O (-)

H (+)   H (+)

O (-)

Pure water is made up of molecules, each having two hydrogen and one oxygen atom. Since oxygen has a molecular weight of 16, and each hydrogen atom has a molecular weight of 1, each water molecule has a molecular weight of 18. Ordinarily, a substance with such a light molecular weight would exist, at normal temperatures, as a gas. For example, oxygen normally exists in nature as an atmospheric gas, yet each molecule contains two oxygen atoms for a molecular weight of $16 + 16 = 32$, nearly double that of water.

Water exists as a liquid because a water molecule is not linear. Instead, each hydrogen atom is offset to the same side of the oxygen atom. Since hydrogen atoms are positive in nature, and oxygen atoms are negative, each molecule becomes bipolar, meaning that there is a positive and a negative side to the molecule. Positive and negative attract, so the positive side of one molecule is attracted to and "sticks" to the negative side of another. This attraction results in a very strong bond, and is the reason water is such a very "light" molecular weight liquid instead of a gas. It is also the reason water is such an aggressive and universal solvent. In fact, its ability to dissolve substances is essential to all plant and animal life, especially in providing the elements needed for the energy-fixing and life-sustaining process of photosynthesis.

Water is more properly seen as a solution, with a composition that varies with the qualities of the air, soil, and rock it has traveled through—and also with the nature of the pollutants that have been introduced into it. In other words, much like human beings, water changes its character based on its experiences. If it is exposed to oxygen and light, it becomes pure and clean and supportive of life. If it is exposed to toxins, it becomes polluted and cannot support life.

Freshwater is distinguished by its relatively low content of total dissolved solids (TDS), as measured in parts per million (ppm). In general, water under 1,000 ppm TDS is considered fresh, though half that (500 TDS) is considered desirable for drinking water. By contrast, sea water has 10,000 to 36,000 ppm of TDS, with over 75% of that in the form of sodium chloride (NaCl). Water between 1,000 and 10,000 ppm of TDS is called brackish, such as the water that exists in mangrove swamps or tidal areas. Water above 36,000 ppm of TDS is known as brine.

In addition to water in the atmosphere, freshwater available for use by humans exists as either surface water or groundwater. Surface water is the runoff draining from the highest land areas, in the form of springs, streams, and rivers, to lower areas, where it collects as lakes, and finally, to the ocean. The area being drained is known as a watershed. Groundwater is normally freshwater that percolates into the ground and is trapped and held in soil and rock formations known as aquifers.

THE NATURE OF GROUNDWATER

Groundwater is the nation's most precious natural resource, accounting for almost 50% of the nation's public water supplies. Over 100 million Americans depend on it for drinking, bathing, cooking, and other domestic uses. American agriculture also depends heavily upon groundwater, as it accounts for over 40% of all water used in irrigation. The EPA estimates that Americans annually withdraw 40 trillion gallons of water from the ground, and the rate of use is increasing at 25% per decade.

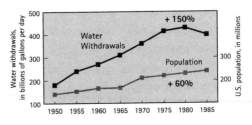

U.S. Water Withdrawal* and Population Trends
* Surface water and groundwater
Source: ZPG and Solley, Merk, and Pearce, 1988

Groundwater is the source of wells and springs. It is formed when precipitation, in the form of rain or snow-melt, infiltrates the ground and moves through the soil until it reaches a zone of rock, where it fills the spaces in the rock formation. If water becomes trapped under pressure between two layers and finds its way to the surface, it is known as artesian water. Water-bearing layers of permeable soil

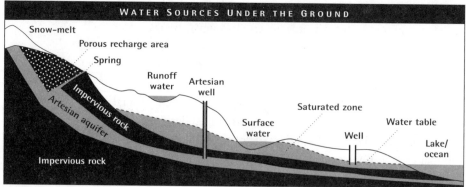

Snow-melt

Porous recharge area

Spring

Runoff water

Artesian well

Impervious rock

Artesian aquifer

Saturated zone

Surface water

Well

Water table

Lake/ocean

Impervious rock

Adapted from *The Home Water Supply* by Stu Campbell, Charlotte, Vermont, 1983.

and rock are called saturated zones. If these zones are large enough to yield significant quantities of water when pumped, they are called aquifers. Aquifers may be located from a few to several thousand feet below the earth's surface. Some areas may contain several aquifers at different depths; some areas have no aquifers.

Groundwater moves very slowly through an aquifer, usually only a few feet a month, if that. Cracks, fissures, and wells greatly influence this movement. The depth at which water may be found in the ground is called the water table. In areas of extensive well-pumping, or during prolonged dry periods, groundwater actually drains out of the soil and the water table drops. This means the depth from which water must be pumped increases.

If groundwater supplies are depleted faster than they can be recharged by the hydrological cycle, the aquifer gradually begins to become depleted, a process variously called water mining, drawing down, or overdraft. If the situation persists long enough, it can lead to subsidence of the land above it. The result can be shifts that damage roads, bridges, and buildings, and cause dramatic sinkhole collapses. Subsidence can cause so much contraction in the volume of the aquifer that it can never regain its former water-holding capacity, even if it is recharged. When that happens, the aquifer changes from a renewable to a declining resource.

Major aquifers have formed over thousands of years. They are cool, dark, and abiotic, which means they do not readily support the

In Norwich, England, groundwater was contaminated with whale oil in 1815. Water drawn from wells today still contains those 180-year-old poisons.

growth of bacteria capable of breaking down pollutants into simpler, less toxic substances. Because water flows so slowly through aquifers, contaminants may be stored for hundreds of thousands of years. What's worse, contamination may not be discovered until decades after it begins, and the slow movement of water often makes locating the source of contamination extremely difficult, if not impossible. For example, in Norwich, England, groundwater was contaminated with whale oil in 1815. Water drawn from wells today still contains those 180-year-old poisons. In Bellevue, Ohio, public and private wastes were dumped into sinkholes and wells beginning in 1872. Today, over 120 years later, those wastes show up in wells drilled over a 75-square-mile area. When pumped, some wells still disgorge raw sewage from that era, including unde-composed toilet tissue.

THE JOURNEY OF A WATER MOLECULE	
Average time a water molecule stays in various levels of the water cycle	
Location	**Residence Time**
Atmosphere	9 days
Rivers	2 weeks
Soil moisture	2 weeks to 1 year
Large lakes	10 years
Underground water at shallow depth	10's to 100's of years
Ocean water to a depth of 55 yards	120 years
Oceans	3,000 to 5,000 years
Underground water at depth	10,000 to 100,000+ years
Antarctic ice cap	10,000 to 100,000+ years

Source: Ehrlich, as cited in United Nations Statistical Commission and Economic Commission for Europe. *The Environment in Europe and North America: Annotated Statistics 1992* (United Nations, New York, 1992). Table II-2.4.6, p.215.

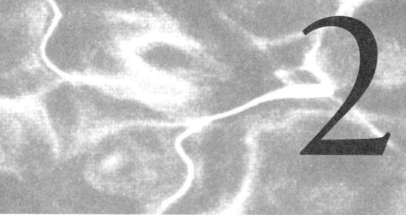

2

In New Mexico,
the Pecos River is so
polluted from mine
tailing contamination
upstream that residents
are advised not to eat
fish from this once
pristine river.

POLLUTION

WATER POLLUTION IS DEFINED as the presence of unwanted substances beyond levels considered acceptable for health or aesthetics. These may include organic (living or once-living) matter, heavy metals, minerals, sediment, biological contaminants, and toxic chemicals.

SOURCES OF POLLUTION

Some contaminants come from natural sources, including radon gas and a variety of salts and minerals which leach into water from the ground. Human activities are responsible for the increasing presence of these and many other contaminants. The major types of water pollution include the following point (coming from a specific location) and non-point (coming from a widespread area) sources:

MINING Mining presents a major threat to groundwater. These point-source activities release toxic heavy metals such as mercury, cadmium, and lead, and sometimes radioactive material like uranium, into the environment. They can also cause acidity and erosion with its resulting sedimentation. Mining can destroy underlying aquifers, change local flow patterns, and contaminate surface waters, as cyanide, heavy metals, and other poisons leach into local rivers and streams. In New Mexico, the Pecos River is so polluted from mine tailing contamination upstream that residents are advised not to eat fish from this once pristine river, or drink water from wells located along its course.

DRILLING FOR OIL AND NATURAL GAS Recently banned in most states, the use of brine pits to dispose of the saline by-products of drilling is a major contributor to groundwater contamination. Texas has reported over 23,000 cases of groundwater and surface water contamination linked to petroleum exploration. Since drilling activities usually occur in sparsely populated areas, their impact, though potentially large, does not receive much public attention. Much waste is totally unregulated, and those regulations which are in place are often poorly, if ever, enforced.

AGRICULTURE AND FORESTRY   These practices are especially problematic where pesticides, which leach into groundwater supplies, and poor farming practices, which encourage soil erosion and sedimentation, are used. Factory farms (feed lots, poultry-raising operations, etc.) are major sources of pollution because of the large and concentrated amounts of animal wastes they produce. Deforestation causes runoff. Instead of permeating soil and recharging aquifers, rainwater erodes hillsides and clogs streams and rivers with silt. Downstream, wetlands become choked with this silt, which destroys wildlife habitat and upsets delicately balanced ecosystems that depend on clean water to support life.

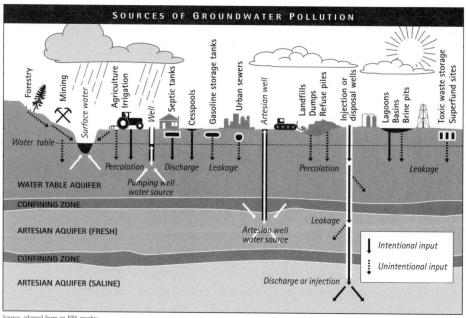

Source: adapted from an EPA graphic.

CITY DRAINAGE SYSTEMS   Fertilizer runoff from lawns, and toxic chemical runoff from streets, are major sources of pollution. These drainage systems often empty into nearby rivers or other bodies of water. In a storm, anything that has accumulated on the street—oil, soot, metals, animal wastes, dirt, litter—is washed into the receiving waters. Residents and business owners who dump used motor oil and other toxic substances into storm drains make matters worse. So far, these non-point sources are regulated poorly, if at all.

LANDFILLS Landfills and open dumps are major point sources of pollutants, and are one of the greatest threats to groundwater supplies. The EPA estimates that there may be as many as 75,000 industrial landfills and 15,000 municipal landfills. American industry generates over 80 million pounds of hazardous waste each year. A large percentage of this waste is dumped in landfills, and only a small percentage is dumped in a way that is considered environmentally secure. A 1979 study, conducted by the Oversight and Investigations Subcommittee of the House Committee on Interstate and Foreign Commerce, found that between 1950 and 1979, the top 14% of the nation's chemical companies, owning or operating 1,600 facilities, dumped 1.5 trillion pounds of industrial wastes at over 3,600 sites. A 1977 study of 50 sites reported that, at 40 of the sites, organic chemical contamination was found in the groundwater; and at 43 of the sites, migration to the underlying aquifer of at least one hazardous chemical was detected. Inorganic chemical contamination of groundwater at 26 of the sites exceeded EPA safety levels for heavy metals in drinking water.

> **Landfills are one of the greatest threats to groundwater supplies. A 1979 study found that between 1950 and 1979, the top 14% of the nation's chemical companies dumped 1.5 trillion pounds of industrial wastes over 3,600 sites.**

Improper hazardous waste disposal is not the only threat posed by landfills. Precipitation falling on landfills can filter through their contents and percolate into the ground below. As this water passes through, it can leach out heavy metals, organic chemicals, and other potentially dangerous compounds contained in garbage at the site. This leachate finds its way into groundwater below. Long after dumping at a landfill has ended, the leaching action continues. Unless measures are taken to stop the formation of leachate, or drain it away safely, inactive landfills will continue to threaten groundwater supplies.

WATER AND SEWAGE TREATMENT PLANTS Many older cities have combined sewer systems, in which storm drains empty into the same sewer lines that carry wastes from homes and businesses. In heavy rainstorms these systems often carry large amounts of runoff, far more than most sewage treatment plants can handle. Faced with this sudden surge, plant operators have to bypass treatment, allowing mixtures of sewage and runoff to flow into rivers or bays. The resulting discharge can cause severe pollution.

DOMESTIC SEPTIC TANKS AND CESSPOOLS In its reports to Congress on waste-disposal effects on groundwater, the EPA estimated that 29% of the American public rely on this form of sewage disposal, including an estimated 20 million single-family households. These systems pump into the ground excessive concentrations of nitrates, phosphates, viruses, and microorganisms from natural wastes. Hazardous wastes such as used automotive oil and antifreeze, batteries, paint removers and solvents, and lawn and garden products (especially pesticides), reach the water supply through cesspools and septic systems. Areas with large numbers of septic tanks and cesspool disposal systems may suffer significant groundwater contamination. In Hawaii, the practice of dumping domestic waste into underground lava tubes has resulted in certain coastal areas becoming polluted and is the likely cause, in certain areas, of a dramatic rise in poisoning from the eating of reef fish.

UNDERGROUND GASOLINE STORAGE TANKS There are hundreds of thousands of underground gasoline storage tanks throughout the nation. Most of the tanks installed during the petroleum boom of the 1950s and 60s are made of steel, and along with their pipe connections were designed to last only 20 years or so. Predictably, many of these now leak significant amounts of gasoline into underground water supplies. Once gasoline has leached into groundwater, it is difficult, if not impossible, to completely remove. Gasoline contamination of an aquifer may render it unusable for decades.

UNDERSTANDING THE POLLUTION PROBLEM

Some people claim that water pollution is an exaggerated problem because less than five percent of America's 35 quadrillion gallons of economically producible groundwater is now known to be polluted. While the statistic is true, the conclusion is faulty. A large percentage of that unpolluted water lies under areas that are either sparsely populated, unsuitable for widespread human activity, or uninhabitable.

Where population is concentrated and human activity in the form of industry and agriculture is prevalent, pollution affects a large percentage of vital groundwater supplies. Not only is groundwater threatened, but surface waters such as streams, rivers, and lakes, and ocean waters off highly populated coastal areas, are dangerously polluted as well. Nature's hydrological cycle strives to replenish the supply, but civilization is polluting this water at a rate that far outstrips nature's ability to cope. Domestic and industrial sewage, chlorine and aluminum additives, industrial chemicals, acid rain, agricultural fertilizers, pesticides and herbicides, and toxic waste dumping, just to name a few, are polluting the precious supply of fresh water which we need for drinking, bathing, and cooking. Our water is becoming a breeding ground for ever more potent bacteria and viruses, and a repository for other toxic contaminants.

Synthetic organic chemicals pose a special threat. Many of these are potent toxic chemicals that cannot be broken down by subsurface microorganisms or filtered out by soils before they reach groundwater supplies. Chlorinated hydrocarbons like pesticides, aldrin/dieldrin, chlordane/heptachlor, kepone, mirex, trichloroethylene (TCE), tetrachloroethylene (PCE)—also known as perchloroethylene—and carbon tetrachloride, are especially difficult. Huge volumes of these chemicals are produced, used, and disposed of each year. Not surprisingly, samples of well water from all over the country show how extensively these compounds have contaminated groundwater supplies.

As a result, where human activity and populations are significant, natural sources of absolutely pure water, once common, are now increasingly rare. What's worse, that which is available is rapidly deteriorating in quality.

> The hydrological cycle strives to replenish the supply, but civilization is polluting groundwater at a rate that far outstrips nature's ability to cope. Our water is becoming a breeding ground for ever more potent bacteria and viruses, and a repository for other toxic contaminants.

3

*The number of
new and different
contaminants released
into our water systems
has grown faster
than the ability of the
government, science,
or citizen's groups to
keep up.*

GOVERNMENT ATTEMPTS AT REGULATION

THE UNITED STATES USES about 350 billion gallons of fresh water a day, or about 1,400 gallons per person. That's more than any other industrialized country, and many times more than any developing nation. About 41% is used for irrigated agriculture; another 38% is used to cool electric power generating plants; about 11% is used by industry, and about 10% is used for public tap water supplies.

Since all natural water supplies contain some dissolved substances, the issue is not whether or not water is pure, but which substances at what level pose a threat to the environment. These questions provide ample fuel for hot debate between business interests, environmentalists, health practitioners, and government. The federal government has attempted to deal with the problem by passing legislation such as the Clean Water, Safe Drinking Water, Surface Mining Control and Reclamation, and Resource Conservation and Recovery Acts. Each Act is designed to deal with some aspect of the pollution problem.

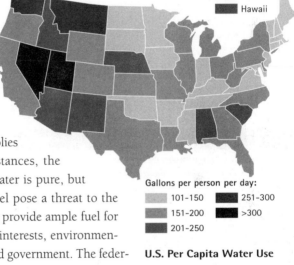

Gallons per person per day:
101–150   251–300
151–200   >300
201–250

U.S. Per Capita Water Use
Source: American Water Works Association

The Environmental Protection Agency (EPA), under 1986 amendments to the Safe Drinking Water Act of 1974, has been charged with setting and enforcing quality standards for the 10% of freshwater supplies devoted to public tap water. Under the law, the EPA sets two standards for each pollutant. The first is called a Maximum Contaminant Level Goal (MCLG). It is defined as the level that is not expected to cause adverse health effects over a lifetime of exposure, and must include a safety margin. There is no penalty for violation of this standard. It is simply a guideline based on what is supposed to be the best current research and scientific knowledge.

The second standard is enforceable. It is called the Maximum Contaminant Level (MCL) and must be set as close to the MCLG as possible, limited only by cost and technology. The MCLs most important to health are called Primary Drinking Water Standards. Those that relate more to aesthetics, including characteristics such as color, taste, odor, hardness, salinity, pH, and turbidity (murkiness), are covered by the Secondary Drinking Water Standards.

There are hundreds of pollutants found in drinking water but the EPA is required to set standards for only 60 of them. According to the EPA, nearly half of all municipal water suppliers annually violate federal health standards.

The problem is that the number of new and different contaminants released into our water systems has grown faster than the ability of government, science, or citizen's groups to keep up. There are hundreds of pollutants found in drinking water, but the EPA is required to set standards for only 60 of them. Alarmingly, even those standards are *routinely* violated. According to the EPA, nearly half of all municipal water suppliers annually violate federal health standards. Considering how these standards are developed and enforced, that record is cause for concern.

U.S. DRINKING WATER DISEASE OUTBREAKS				
	OUTBREAKS		TOTAL	
	1991	1992	Outbreaks	Cases
AGI*	12	11	23	13,367
Giardia	2	2	4	123
Cryptosporidium	1	2	3	3,551
Hepatitis A	0	1	1	10
Shigella Sonnei	0	1	1	150
Nitrate**	0	1	1	1
Fluoride**	0	1	1	262
Total	**15**	**19**	**34**	**17,464**

* Acute gastrointestinal illness of unknown etiology
** Not microbiological contaminants, but can cause disease

Source: Centers for Disease Control and Prevention and the U.S. Environmental Protection Agency collaborative surveillance program.

1993 outbreaks of Cryptosporidium in Milwaukee, and Las Vegas, Nevada, left thousands sick and over a dozen dead. A similar outbreak in Albuquerque in May of 1994 led to television newscast warnings for people with weakened immune systems to boil their water. Early in 1994, dangerous E. Coli bacteria were found in New York City water. At about the same time, Washington, D. C. residents were being told to boil their water for three days because bloodworms (chlorine-resistant nematodes) were found in the water system.

President Clinton, in his State of the Union address, vowed to revise the Safe Drinking Water Act of 1974 (the Act). When the Act was last re-authorized in 1986, it mandated the EPA to set national standards for contaminants in water supplies, and to see that water systems use the "best available technology" to meet these standards. Since then, Congress and the EPA have received a tidal wave of complaints from municipalities who argue they can't afford to install the technology to meet those standards, much less the new standards which are being proposed by Clinton and others in several re-authorization bills. Many of them are lobbying hard to have the maximum contaminant levels (MCLs) raised to make them easier and less expensive to meet, rather than further reduced to levels neces-

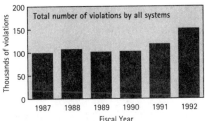

U.S. Drinking Water Violation Trends
Source: FRDS 9/93

sary to protect your health. These municipalities have received a lot of support from the new Republican majority in Congress, many members of which have promised to weaken existing regulations.

The sad fact is that city infrastructures are decaying while the amount and types of pollution are increasing dramatically. States complain that there isn't enough money to maintain our highways or schools, much less rebuild obsolete water treatment facilities, most of which are running on archaic technology developed prior to or during World War I.

While municipalities struggle to maintain outdated technology, every year 18 billion pounds of new pollutants and chemicals are released by industry into the atmosphere, soil, and groundwater. Over 70,000 different chemical compounds are now in use by industry, agriculture, and private citizens; 5,000 new and unproved chemical compounds are being introduced into the environment every year. At least 700 of these chemicals have been regularly found in America's drinking water, but, as we have noted, the EPA has set safety standards for only 60 of them.

According to the Natural Resources Defense Council (see "Resources" at the back of this book), the public is being deliberately misled about the seriousness of the problem. They say that many public water suppliers know of contamination problems and yet, in a direct break with their public trust responsibilities and with the law, fail to notify their customers. They even go so far as to say that some water suppliers deliberately falsify their water test results, potentially threatening the health of the public.

Their analysis reveals that there is a severe lack of investment in the nation's water treatment systems. Most still rely upon obsolete technology (developed in the early 1900s) that fails to remove many dangerous contaminants. They also say that many of the state programs for controlling drinking water problems are in disarray, enfeebled by over a decade of starvation for resources and lack of

To verify their conclusions, the Natural Resources Defense Council points to the following facts:

1 There is a breakdown in compliance with the Safe Drinking Water Act

According to the EPA's records, in 1991-1992, the nation's water systems committed over 250,000 violations of the Safe Drinking Water Act, affecting more than 100 million Americans. Roughly 10%, or over 25,000 of these violations, were serious, because they were violations of EPA's fundamental health standards, the Maximum Contaminant Levels (MCLs).

2 Some water suppliers practice under-reporting and outright deceit in reporting drinking water contamination

The General Accounting Office (GAO) and internal EPA documents show clear evidence of under-reporting of violations by water systems and by many states. For example, many states supplied no monitoring data for entire classes of contaminants. Also, according to GAO and EPA records, many water suppliers file, or are suspected of having filed, falsified reports about the level of contamination in their water. Significantly, neither the EPA nor many states have developed effective programs to root out falsification and to prosecute the violators.

3 There is a widespread failure to enforce

Despite widespread noncompliance, and even outright falsification, the EPA's data shows a clear pattern of failure to enforce. In the face of over 250,000 violations, states took just over 2,600 formal final enforcement actions, and the EPA took about 600. Thus, only a tiny percent of the violations were ever subject to penalties.

4 Water suppliers routinely fail to effectively notify the public of drinking water contamination

Although the law says that drinking water suppliers must notify their customers when their water is contaminated, or when the system has violated EPA rules by failing to test water for contamination, in the majority of cases the water suppliers do not provide this notice to the public. If notice is provided, it is often done in a manner calculated to ensure that few, if any, customers are actually informed of, or understand, the problem. For example, many water systems typically place a notice in the "Legal Notices" section of the newspaper, virtually assuring that the public is never made aware of the problem.

adequate EPA oversight.

While the EPA has cataloged over 250,000 serious unsafe water violations affecting over 120 million people on public water systems, people on private wells fare even worse. EPA surveys of private wells show that fully two-thirds violate at least one Safe Drinking Water Act Standard.

Put simply, *you can't depend on the government to fix the problem!* The public health implications of the nation's drinking water problem are staggering. As we have seen, contaminants that are regulated are not adequately controlled under EPA rules, because some EPA standards are far too low, and violations are subject to weak or nonexistent enforcement. In addition, many important contaminants are completely unregulated.

Some citizens who are becoming aware of pollution problems are working to force government to regulate many a reluctant industry into expensive changes. The EPA, responding to citizen pressure, has repeatedly revised the enforceable standards that municipalities must meet in their water treatment facilities. Public water suppliers are fighting back. They say that the changes demanded by consumers are too expensive, and will mean that either taxes or water fees have to be raised, other important programs reduced or eliminated, or both. Politicians hate that sort of problem, so there is great resistance to raising the standards or furthering programs to educate the public about the matter.

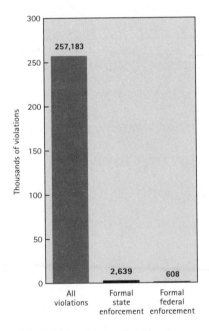

U.S. Drinking Water Violations and Formal Enforcement Actions, 1991–1992

Source: FRDS 19, FRDS 17 (8/93)

4

Don't assume that just because the water coming out of your tap is tested that it is safe. The EPA has cataloged over a quarter million unsafe water violations affecting over 120 million people on public water systems.

PUBLIC WATER TREATMENT METHODS

INCREASING POPULATIONS HAVE MEANT an increasing demand for potable water for human consumption. With demand continuing to outstrip a diminishing supply, the situation has left no alternative but to use even more chemicals in ever higher concentrations to treat water in an effort to make it usable. This is a fact which causes a great concern to knowledgeable researchers in governmental, health and environmental organizations.

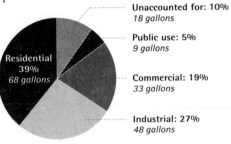

Unaccounted for: 10%
18 gallons

Public use: 5%
9 gallons

Residential
39%
68 gallons

Commercial: 19%
33 gallons

Industrial: 27%
48 gallons

Treated Water Distribution in the U.S.
Of 176 gallons per person per day treated
Source: American Water Works Association

Public treatment systems vary widely, but an overview of common steps would include:

COAGULATION Chemicals are added to create small jelly-like particles, called floc, which collect dirt and other solids suspended in the water.

FLOCCULATION Water is gently mixed so floc particles clump together.

SETTLING Larger floc particles and sediment are slowed so they fall to the bottom, creating sludge.

FILTRATION Water is passed through a granular medium such as sand or crushed hard coal.

DISINFECTION Chlorine or other chemicals are added to the water to kill potentially disease-causing bacteria and other microorganisms.

CORROSION CONTROL Chemicals like quicklime are added to reduce the water's acidity, to prevent corrosion in water pipes and the consequent leaching of dangerous metals into the water supply.

FLUORIDATION Fluoride is added for the questionable purpose of preventing tooth decay.

AERATION Air is passed through water so volatile organic compounds will evaporate and be removed.

While modern treatment technology, where employed, has clearly resulted in a reduction or elimination of the dreaded plagues of cholera and typhoid, it has created a whole new set of health considerations. For example, the large amounts of aluminum used in the coagulation and flocculation processes has been linked to the age-related diseases of Alzheimer's and senile dementia. Chlorine disinfection processes are becoming increasingly linked to the dramatic rise in heart disease and cancer. New information about fluoridation raises questions about possible links to arthritis and cancer.

Flow Diagram of a Typical Water Treatment Plant

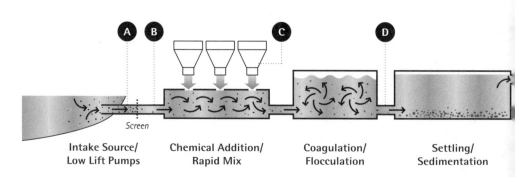

| Intake Source/
Low Lift Pumps | Chemical Addition/
Rapid Mix | Coagulation/
Flocculation | Settling/
Sedimentation |

APPLICATION POINTS FOR CHEMICALS

Chemical Category	Possible points of application							
	A	B	C	D*	E	F	G	H
Algicide	●				●			
Disinfectant		●	●		●	●	●	●
Activated carbon		●	●		●	●		
Coagulants		●	●					
Coagulation aids		●	●		●			
Alkali:								
for flocculation			●					
for corrosion control					●			
for softening			●					
Acidifier			●		●			
Fluoride					●			
Cupric–chloramine					●			
Dechlorinating agent					●			●

*With solids-contact reactors, point C same as point D.

Source: *Water Treatment Plant Design*, American Water Works Association, New York, 1969.

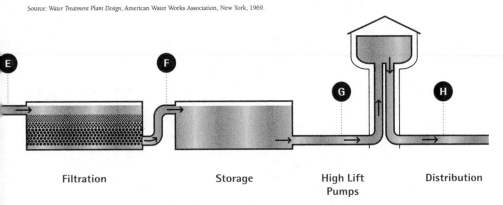

Filtration   Storage   High Lift Pumps   Distribution

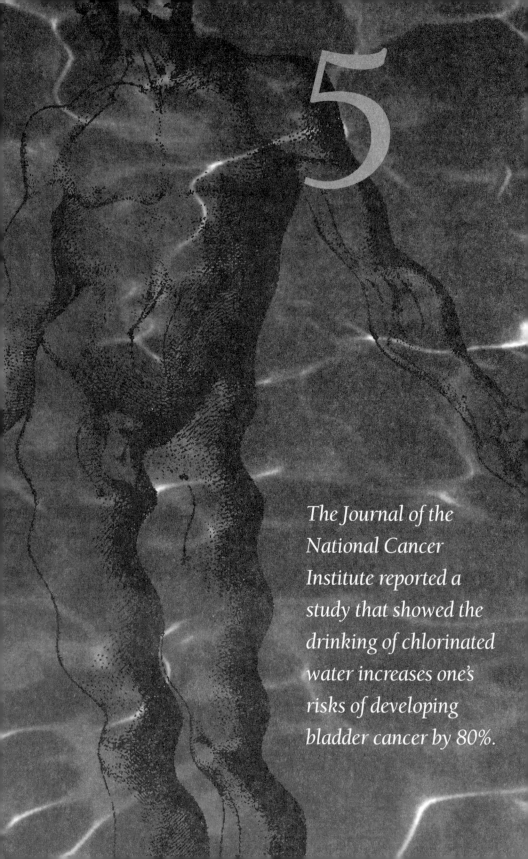

5

The Journal of the
National Cancer
Institute reported a
study that showed the
drinking of chlorinated
water increases one's
risks of developing
bladder cancer by 80%.

DRINKING WATER AND YOUR HEALTH

THE ROLE OF WATER in metabolism, assimilation, regulation of body temperatures, and nourishment of body tissues is critical. Breathing, digestion, elimination, glandular activities, heat dissipation, and secretion are properly performed only when enough water is present in the body.

To remain healthy, a normal adult must consume approximately 64 ounces of pure water a day. That's one 8-ounce glass every other waking hour! Most Americans consume far less; instead, the consumption of water has been replaced by caffeinated beverages such as coffee and tea, soda pop, and alcoholic beverages. Although these beverages contain water, they are also diuretics. That means they cause the body to lose water, creating more dehydration, rather than less.

> To remain healthy, a normal adult must consume approximately 64 ounces of pure water a day. That's one 8-ounce glass every other waking hour!

Those who short themselves by not drinking enough pure water predispose themselves to illness. Unintentional chronic dehydration is at the root of many serious diseases, including asthma, renal dysfunction, endocrine system and adrenal fatigue, high blood pressure and other cardiovascular problems, arthritis, ulcers, pancreatitis, digestive difficulties, and lower back pain.[2]

THE PROBLEM OF DEHYDRATION

Most Americans are unaware of the health problems caused by dehydration, because the orthodox medical establishment in this country has been very slow to acknowledge the role of nutritional imbalances, including chronic dehydration, in disease; and slow to recognize the role of adequate nutrition and clean drinking water in overall public health. It is more profitable to perform expensive tests, surgery, and prescribe dangerous pharmaceuticals than it is to advise a patient to drink more clean water and improve diet. As a result, the medical establishment continues to center its treatment options around the prescription of chemicals and/or surgery, while cooperating with the pharamaceutical industry and government to do everything possible to restrict access to alternative methods of healing. Things are changing, but the average doctor graduating from medical school has had less training in the principles of nutrition than the average counselor at a weight loss clinic. As a result, health care is getting more expensive while Americans are getting sicker.

The handwriting is on the wall. Sooner or later, attitudes must change. Those at the forefront of nutritional research are quietly bringing forth information that promises to revolutionize medicine as we have come to know it.

Simple nutritional therapies are working miracles where traditional medicine has failed. In all of these therapies, drinking an adequate supply of pure water is one of the first steps to the restoration of health. To understand the difference this can make, consider the difference between how allopathic (traditional western medicine) and naturopathic (natural medicine) practitioners handle a common disease like asthma.

Asthma occurs when the bronchioles of the lungs constrict, causing breathing to become dangerously difficult. In severe cases it is life threatening. A traditional doctor will likely prescribe some form of antihistamine or other bronchodilator to alleviate symptoms. But that treats the symptom, not the cause.

Naturopaths, trained in the role of proper nutrition, take a deeper look. They know that the body loses water mainly through exhaled breath, urination, perspiration and elimination of fecal matter. Since the body can lose up to a gallon of water a day, that water must be replaced. If it is not, the body begins to conserve. One way it does this is to release histamine into the lungs, thus constricting the bronchioles in order to stem water loss. This results in the symptoms of asthma.

The proof that the majority of asthma cases are really misdiagnosed cases of chronic dehydration, is that as soon as water balance in the tissues is restored, symptoms often cease in a large majority of patients. Some studies report as many as 94% of asthma patients find dramatic relief of symptoms by drinking large volumes of water at the first sign of breathing difficulty. Many naturopathic practitioners report that when their patients maintain appropriate water consumption, and glandular function is restored, symptoms often

Heart disease

Asthma

Ulcer

High blood pressure

Arthritis

Gastrointestinal illness

Health Problems Associated with Chronic Dehydration

disappear completely.

Naturopaths also know that dehydration causes edema and high blood pressure. A water-starved body instructs the kidneys to conserve sodium, which concentrates in the body's tissues, thereby retaining water otherwise used for perspiration and urination. The adrenals release hormones which cause arteries to constrict so that fluids are retained. This causes blood pressure to go up and is the reason many asthmatics have high blood pressure.

As dehydration continues, the body eventually loses too much water to function properly. Great stress is placed on the endocrine system, and hormone balance is disrupted. Lymphatic fluid becomes thicker, resulting in stress to the immune system and a general loss of tissue oxygenation.

Many other common diseases are really misdiagnosed cases of dehydration. For example, the mucous that lines the stomach to protect it from hydrochloric acid is 98% water. When this mucous barrier has too little water, it sloughs off and fails to protect the stomach cells from stomach acid. Ulcers can be the result.

Lack of water means the pancreas cannot produce the sodium bicarbonate solution necessary to protect the duodenum from stomach acid. The resultant severe cramping is often misdiagnosed as pancreatitis.

The synovial fluid which lubricates the joints is primarily water. Lack of synovial fluid, combined with increased levels of histamine in the tissues (remember how the body releases histamine to conserve water by constricting bronchioles in the lungs), causes joint inflammation that eventually turns into arthritis.

When chronic dehydration is not recognized as the underlying cause of the problems, the patient may be treated with inappropriate drugs such as antihistamines, steroids, and antibiotics. All of these have dangerous side effects, and can cause further damage to important organs and delicate systems.

Doctors and patients alike have a hard time understanding the problem of dehydration because the patient doesn't feel thirsty. But it's erroneous to assume that anyone who is dehydrated feels thirsty. That's because thirst is a biological response subject to the influence of conditioning. Humans instinctively dislike the taste of the chlorinated chemical beverage that passes for water

Humans instinctively dislike the taste of the chlorinated chemical beverage that passes for water in most places. As a result, they learn to avoid drinking it. Gradually, people who do this turn off their thirst response and recognize it only when it is severe. That's why those who drink water only when they are thirsty are usually dehydrated!

in most places. As a result, they learn to avoid drinking it, substituting flavored diuretic beverages like coffee, tea, soda and beer. This causes even more water loss! Gradually, people who do this learn to turn off their thirst response and recognize thirst only when it is severe. That's why those who drink water only when they feel thirsty are usually dehydrated!

THE CONTAMINATION OF PUBLIC WATER SUPPLIES

Dehydration is only one health problem associated with water. Of greater concern is the quality of the water Americans are drinking. Outbreaks of serious disease caused by water-borne parasites and toxic pollution are increasing at an alarming rate. So, not only do we not drink enough water, we are unaware of problems with the water we do drink. That's a big mistake. William K. Reilly, Administrator of the Environmental Protection Agency (EPA) under the Bush Administration, as well as the scientists who make up the EPA's Science Advisory Board, all agree that drinking water contamination in the United States ranks as one of the top public health risks.[3]

> The government is finally being forced to admit that outdated water treatment facilities throughout the country are incapable of coping with the flood of environmental toxins that are permeating the nation's water supply.

One in six people drink water contaminated by excessive amounts of lead. In the early summer, half the rivers and streams in America's corn belt are laced with dangerous levels of pesticides. Microbes in tap water are thought to be responsible for at least one in three cases of gastrointestinal illness.

According to the Natural Resources Defense Council, nearly a million people a year are affected by drinking contaminated water. Their report concludes: "With the information now available, our trust in our drinking water supply can and will be shaken. We must think before we drink."[4]

The government is finally being forced to admit that outdated water treatment facilities throughout the country are incapable of coping with the flood of environmental toxins that are permeating the nation's water supply.

Not only are old technologies outdated, some are actually dangerous. One example is the widespread use of chlorine as the primary means of disinfecting water. On the one hand, it is ineffective against an increasing variety of dangerous pathogens which are finding their way into the nation's water supply. On the other, new research indicates that the ingestion of chlorinated water may be a primary cause of cancer and heart disease, two of our nation's leading killers (see *Chlorine* in Chapter Seven).

Another example is the widespread use of fluoridation to prevent tooth decay. Many knowledgeable critics claim that fluoridation presents an unacceptable risk to health because of its effect on the body's immune system, as well as its potential genetic, mutagenic, and carcinogenic effects. Even the Environmental Protection Agency has called for further studies to determine the extent of the risk.

Statistics compiled by the National Centers for Disease Control and Prevention confirm that nearly a million people are sickened, 350,000 are exposed to dangerous levels of carcinogens *other* than chlorine, and 900 people die from drinking contaminated water each year. Yet, with few exceptions, drinking water contamination has received relatively little national or local media attention. Few people are aware that, according to the most recent published review of studies of actual cancer cases in the United States, a single class of drinking water contaminants (known as trihalomethanes and their chemical cousins) is associated with 10,700 or more bladder and rectal cancers per year—about thirty cancers per day.[5] To put it into perspective, that is twice the number that die from fires, and more than are killed each year by handguns.

> "Few things are as insidious as bad water. It's dangerous for you and your children, but you usually can't tell if you have it. And if you do, you may not be able to find out where the problems are coming from. Water can carry some of our most serious diseases—typhoid, dysentery, hepatitis—yet still look clear in the glass. We may do battle over how we get our water and develop it, but we fear for its quality."
>
> —*National Geographic Magazine,* Special Water Edition, November 1993.

As surprising as those statistics are, few people are aware of the risk of trihalomethanes. Even fewer know of the many other common chemical and biological drinking water contaminants that are known to cause or predispose over 125 million Americans to serious illness such as cancer, immune system dysfunction and heart disease.

6

"With the information now available, our trust in our drinking water can and will be shaken. We must think before we drink."

— National Resources Defense Council report

THE GROWING DEMAND FOR PURE WATER

RAPID INDUSTRY GROWTH

Increasing awareness of the dangers presented by drinking water accounts for the growth of the American "point-of-use" water treatment industry (which includes bottled water and treatment devices) into a $7 billion dollar a year business. There are nearly 15 million household filtration systems in place, and surveys show an annual growth in nationwide industry sales of 20% per year. Bottled water is by far the fastest-growing segment of the beverage industry.

At the rate the water treatment industry is growing, by the turn of the century as many American households will own a water treatment device as own a television. The point has not been lost on American businesspeople. Over 500 manufacturers, making thousands of water treatment products, have sprung into existence since 1980. Not surprisingly, in their fervor to capture market share, every manufacturer claims its product is best. As with other industries experiencing rapid growth, some shakeout is inevitable, but the process leaves consumers confused.

Only a few states require manufacturers to test their products to prove they live up to advertised performance ratings. When required, these tests are usually performed by the National Sanitation Foundation (NSF) which provides independent certification that the product meets certain standards. Such certification programs are important, but until people know the nature and scope of water contamination problems and how to interpret performance ratings, such requirements aren't really very helpful. Evaluations such as the ratings prepared by the publishers of *Consumer Reports* help, but they frequently don't tell the whole story. While these are valuable resources, they are no substitute for personal responsibility. In other words, no one is going to help you any more than you help yourself. To help yourself, you must clearly understand the problem before looking for a solution.

1 Take responsibility.

Realize that you are responsible, because you must live with the consequences of your choices. People in government are human and make mistakes just like you. Just like you, the government sometimes avoids problems because they are expensive or difficult to fix. Sometimes things happen at water treatment plants that are beyond the capability of the plant operators to avoid. By taking back responsibility, you create the opportunity to have control over your health. While water quality is a serious concern, it is a problem that every responsible person can easily solve for themselves. All that is needed is awareness and motivation. Modern technology has placed excellent water conditioning devices within the economic reach of nearly every American.

2 Have your water analyzed by a competent professional.

Many companies offer "free" water tests, but only analyze for hardness, in an attempt to convince you to buy a water softener or a water softener coupled with a reverse-osmosis system. While the best of these systems are fine for their purpose, some are costly to buy, costly to maintain, and have significant limitations.

If you are on municipal water, the analysis should include a test for lead, chlorine, fluoride, nitrates, and hardness. Other potential contaminants can be ascertained by contacting your local water company, which is required to disclose the results of their tests to customers who ask for them.

If you are on a private well or water system, the analysis should add tests for pH, iron, sulfates, biological contaminants, herbicides, and pesticides. If your location means that your water could be contaminated with radon, toxic waste from landfills, heavy metals like mercury, cadmium, and arsenic from mine tailings, or leaking fuels and oils from underground storage tanks, it would be very wise to have appropriate tests done for those problems as well.

Don't assume that the water coming out of your tap is safe just because it has been tested. Nitrate levels can fluctuate widely according to season. There are some biological contaminants like Cryptosporidium and Giardia that are not affected by chlorination, and routine tests for these and many other contaminants are not yet required by the EPA or regularly performed by water suppliers.

3 **Once the test results are in, compare what's in your water to the section of this book entitled: "Understanding the Contaminants."**
Pay particular attention to aluminum, arsenic, biological contaminants, chlorine, fluoride, hydrogen sulfide, hydrocarbons, nitrates, and radon, because they are most dangerous to your health. Also pay attention to pH, hardness, manganese and iron. While they are normally not harmful to health, they affect the kind of technology necessary to solve your water problem.

4 **After you understand the danger, if any, of the contaminants found in your drinking water supply, read the section of this book entitled: "Understanding Your Choices."**
If you decide to purchase a filtration system, be clear about the pros and cons of each technology, the cost to purchase, install, and maintain it, and its projected life.

5 **Once you know what you are looking for, shop and compare.**
Because most customers' experience with this industry proves that shopping with a big-name company is no reliable guide to value, be sure to read the section of this book entitled "Shopping Tips."

7

*Every year, 18 billion
pounds of new pollutants
and chemicals are
released by industry
into the atmosphere,
soil and groundwater.*

UNDERSTANDING
THE CONTAMINANTS

WATER CONTAMINANTS CAN BE broken down into two broad categories. The first contains substances that are currently considered relatively harmless to ingest, but may affect water's taste or appearance, or cause damage or staining to plumbing and/or fixtures. The second contains substances that present proven or likely health hazards.

The "cosmetic" contaminants include such minerals as iron, copper, manganese, calcium and magnesium. The more serious health hazard contaminants include chlorine and its by-products, known as disinfection by-products (DBPs); a spectrum of microorganisms called "biological" contaminants; an array of inorganic and organic chemical compounds; a variety of heavy metals; and radioactive materials called radionuclides.

MAJOR NON-TOXIC CONTAMINANTS

ACIDITY   Acidity is usually caused by the presence of mineral acids, salts of strong acids, or free carbon dioxide dissolved in the water. In surface or groundwaters, acidity may come from natural substances or industrial pollution. High levels can corrode plumbing and leach lead and cadmium from solder and other lead-alloy plumbing parts commonly used in well pumps and fixtures. Acidity is a serious problem in Appalachia and certain areas in the West where streams receive acid drainage from coal mines. Other areas of naturally soft (acidic) water exist in the Northeast and Northwest. Acidity is controlled by the addition of soda ash or baking soda to the water supply.

ALKALINITY   Alkalinity is caused by the presence of carbonates, bicarbonates, hydroxides and other dissolved salts. Common over most of the country, and the Southwest in particular, alkalinity is a natural contaminant derived from the atmosphere, the soil, and the solution of carbonate-containing minerals. Alkalinity is not known to be detrimental to humans, but excessive levels can be removed by de-alkalization equipment similar to water softeners. Some alkalinity can be a benefit because it causes a layer of "mineral scale" to coat plumbing parts, preventing lead and other heavy metals from leaching into your drinking water.

COPPER   Copper may exist in natural waters and effluents as a soluble salt or as suspended solids. A small amount is essential for plants and animals, but concentrations exceeding 0.1 mg/liter are toxic to algae, plankton, and some fish. In very young children, a lack of copper can lead to nutritional anemia. Adults require at least 3 mg per day.

For the most part, copper is readily passed from the body in waste, although it does tend to accumulate in the liver. Above 1.0 mg/L, copper can impart an undesirable taste to water. Concentrations of copper in water in excess of 0.05 mg/L are generally the result of pollution, either from industrial or mining wastes, or the corrosion of copper plumbing. In a few cases, it can originate from copper sulfate which is used to control undesirable populations of plankton or algae in lakes, ponds, and reservoirs. Excessive levels of copper can be removed by reverse-osmosis, distillation, ion-exchange and KDF® redox media (see Chapter Eight).

COPPER

Symbol: **Cu**

Small amounts healthy

Bad taste above 1 mg/L

Accumulates in liver

Source: Natural, industry, mining, corroded pipes

Removal: Distillation, reverse-osmosis, KDF®, ion-exchange

HARDNESS   Hardness occurs in conjunction with alkalinity, and enters a water supply when calcium and magnesium salts are dissolved by groundwater. Other di- and trivalent metals have a similar effect, but usually are not present in high enough concentrations to cause problems.

Hardness is commonly measured in grains per gallon (gpg). One gpg = 17 mg/L (milligrams per liter) or 17 ppm (parts per million). Soft water has 0 gpg of hardness. Ideal water has 3 to 7 gpg of hardness. Water is considered very hard at 10 gpg of hardness.

Some hardness is desirable. As said previously, completely soft water tends to be acidic, causing corrosion on the surface of pipes and plumbing fixtures, and leaching lead into drinking water from lead solder and plumbing parts. Also, most hard water tastes better than soft water because of the dissolved minerals.

Unfortunately, many areas of the Southwest have water supplies that measure from 10 to more than 60 gpg of hardness. As levels increase above 10 gpg, hardness becomes an increasingly serious economic problem because of the mineral scale it forms.

Mineral scale occurs when calcium and magnesium salts precipitate out of water, building an amorphous crystalline structure that builds up on the surface of pipes, hot water heaters, plumbing fixtures, appliances, and anything washed in it. At 20 gpg, this buildup of scale can be so severe that hot water heaters may have to be replaced as often as every three years, and the service life of appliances can be cut by half.

Over the life span of a water heater, very hard water will cost much more to heat than soft water. That's because heating the water through the scale build-up on the inside of the hot water heater is like cooking with a ceramic tile placed between the pot and the burner.

If you amortize in the cost of installing two to five new water heaters every ten years, it becomes clear that hard water is expensive. But it doesn't end there. There are other expensive economic costs associated with hardness scale. The harder the water, the higher the cleaning costs. Hardness ions combine with soap to create an insoluble curd. This curd is deposited not only on plumbing fixtures, but on anything exposed to it. Washing will cause skin to become coated with it. Clothes washed in it will become coated as well; then, when clothes go through the dryer, this curd (which has coated individual fabric fibers) becomes hard. As the clothes tumble in the dryer, these fibers break, causing clothes to wear out up to a third faster than clothes washed in soft water.

At 10 gpg of hardness, the average household can use up to four times the amount of soap or detergents as would be required for soft water. To counteract this, most household cleaning products contain numerous chemicals designed to remove hardness so the soap they contain can do its job. These chemicals are then flushed down the sewer, where they can leach into groundwater as phosphates and nitrates, further degrading water quality for whomever lives downstream. Many companies advertise their water-softening products as "environmentally responsible," because they can eliminate the need to use products that contain high levels of phosphates and nitrates.

There is much confusion surrounding the supposed health effects of drinking hard water. Some firms that sell water conditioning products (distillers and reverse-osmosis systems) to remove hardness claim their water is healthier to drink because it is purer. They site the opinions of some doctors who have linked the drinking of very hard water to heart disease, rheumatism, gout, arthritis, and the formation of bladder stones.

Some of this concern comes from a 1974 study comparing death rates of matched groups in Kansas City, Missouri and Kansas City, Kansas.[6] Both cities have the Missouri River as their source of water; but Kansas City, Missouri softens its water before delivery and Kansas City, Kansas does not. The Kansas group, drinking unsoftened water, had a 50% higher death rate from heart disease, and significantly higher blood pressure.

Much was made of these findings, but careful examination of the research data shows that it was not the hardness that was causing the problem, but cadmium. The group drinking untreated water was getting thirteen times the level of cadmium than its Missouri counterpart. Cadmium is a heavy metal removed by water softening technology as well as most other forms of filtration systems, except simple sediment filters.

While some people would like you to believe that soft water is better for you, many studies show that drinking mineralized water is important because it is an excellent source of both the alkaline and trace minerals needed for health.[7] Natural, organic foods are another excellent source.

For any nutrient to be of use to the human body, it must be able to be absorbed by the body. This ease of absorption is called bioavailability. Minerals in water exist as ions, and ionized minerals in food and water are now known to be the most bioavailable form for our bodies. Chelated minerals (like those found in premium brands of vitamin-mineral supplements) are the next most bioavailable, and the least bioavailable are the unchelated minerals like those found in powdered bonemeal or lime. It is not surprising that our bodies derive needed minerals from water. After all, our bodies have evolved over millions of years to be in harmony with their environment. Hard water contains large amounts of ionized calcium, phosphorus, potassium, and magnesium, all essential nutrients and necessary to health.

At 10 gpg of hardness, the average household can use up to four times as much soap and detergents as would be required for soft water.

In 1985, scientists at Oak Ridge National Laboratories decided to test the hypothesis that the calcium and magnesium from hard water could reduce mortality from heart attacks and strokes. While the study failed to prove or disprove that contention, it did find that chlorinated water increases the risk of heart disease because it creates abnormalities in the body's ability to metabolize fats. (This subject is discussed more fully under *Chlorine*, in this chapter.)

Extensive studies confirm that, to the extent hard water provides certain trace organic minerals, it is healthier to drink. Another reason some hardness in the water is desirable is that pure water is a very aggressive solvent; in a highly purified state, it can leach toxins from certain kinds of plastic and other containers used to store it.

No clear evidence exists to link the drinking of hard water to internal health problems of any kind. However, bathing in very hard water does cause dry, itchy, skin and brittle hair. This is partly due to the insoluble soap curd that is deposited on skin and hair and part is due to the harsh detergents and caustic water conditioning chemicals needed to unlock hard water's cleaning power.

> **Surveys show that at 25 gpg, hard water costs the average family over $75 per month. At 50 gpg, the cost soars to over $125 per month.**

If your water has over 10.0 gpg of hardness, an investment in quality water conditioning equipment to reduce the problem of scale will pay for itself over time. The more gpg of hardness, the faster the pay-back. In the West and other areas where hardness can range from 10 to 60 gpg or even more, quality water conditioning equipment becomes an economic necessity. While a monthly payment for such equipment seems expensive, any quality system will actually save money for the family budget. Surveys show that at 25 gpg, hard water costs the average family over $75 per month. At 50 gpg, the cost soars to over $125 per month.

Various technologies exist to remove hardness from water. The most common system utilizes ion exchange resin recharged by standard sodium or potassium salt. Known as water softeners, the best of these work well to remove hardness. The technology is seasoned and reliable, but they can be expensive to install and maintain. If large amounts of iron are present in the water (see the section on iron which follows), some systems are prone to fouling. Another problem is that softeners require the frequent addition of salt to a brine tank, and they add salt to the finished water. When cleaning themselves, the process of backwashing can send a strong stream of brine back into the environment.

Reverse-osmosis (RO) systems remove up to 97% of hardness-causing ions, but domestic units are not practical to treat all the water required for bathing, washing clothes, etc. in a typical household. In fact, most require a softener in-line ahead of them because very hard water can plug up RO membranes.

There are two new interesting technologies for handling hardness problems. The first is a close cousin to reverse-osmosis known as nanofiltration. Like RO, nanofiltration uses membrane technology to separate out hardness-causing ions. However, pore size is considerably larger, permitting a water flow sufficient for domestic needs.

The second is the copper-zinc redox media, marketed under the trade name KDF®, which works by changing the hardness-causing ions from a crystalline into a non-crystalline form. Once changed, they cannot precipitate out of the water to form the amorphous crystalline scale that is one cause of problems associated with hard water.

One limitation with KDF® systems is that while they can reduce the problems associated with lime scale, they don't eliminate it, nor do they do anything to mitigate soap use. If hardness levels are very high, they may not be able to prevent lime scale formation at all, but they are often a better choice than a water softener when hardness levels are moderate (10 to 15 gpg), because they also remove chlorine, heavy metals, and render the water bacteriostatic.

While both nanofiltration and KDF® systems cost more up front, they each have a significant advantage in that they require no salt or other chemicals to work, and add no salt to the water. But, as with water softeners, if excessive levels of sediment or compounds like iron and/or manganese are present, pretreatment may be required to prevent fouling of the medium.

A controversial technology is the use of powerful magnets to change the electrical configuration of the hardness ions so they can't precipitate out as hardness scale. Manufacturers of this technology make claims that have not been backed up with objective proof. Not only does the technology appear to be useless, but when such magnets are installed on boiler systems, operators may avoid routine maintenance, causing damage to the boiler and the need for expensive repairs.[8]

IRON AND MANGANESE Iron-free water is relatively easy and inexpensive to treat, but since iron is one of the most prevalent elements in the earth's crust, most groundwater supplies contain iron in some fashion. If you find reddish stains on your plumbing fixtures, the problem is iron.

When it is dissolved in water, iron is called ferrous iron or "clearwater" iron. Ferrous iron can cause water to have a bloody-metallic taste and odor. Ferrous iron oxidizes on exposure to air to form insoluble, reddish-brown, stain-causing particles of rust. This rust, or ferric iron, can create havoc in plumbing systems, water softeners, and other water-using devices. Colloidal iron is very small particles of oxidized iron suspended in the water. They are usually bound together with other substances. When iron combines with tannins and other organics, complexes are formed that cannot be removed by ion-exchange systems or many types of filtration. If you find reddish stains on your plumbing fixtures, the problem is iron. The presence of certain strains of bacteria that feed on iron further complicates matters. These bacteria create a gelatinous, filamentous sludge capable of clogging valves, plumbing fittings, and water-using appliances, often rendering them useless. Even very small amounts of these bacteria can create a large problem.

The removal of iron can be one of the more difficult tasks in water conditioning. There are several available alternatives. One method is the use of a water softener charged with a special resin designed to handle iron. These units can handle iron up to about 6 ppm (parts per million) depending upon the amount and type of resin used. Water softeners can be used to remove iron only if the water is clear when drawn and no iron bacteria are present.

Water softeners can be used to remove iron only when the water is clear when drawn and no iron bacteria are present. KDF® works in the same way. When the type or amount of iron exceeds the treatment limits of a water softener or KDF® unit, additional treatment becomes necessary.

Catalytic oxidizing filters employ a medium that has been impregnated with various oxides of manganese. As ferrous iron bearing water passes through the filter, iron comes in contact with the medium and oxidizes to form insoluble ferric iron. The resulting rust particles are trapped in the filter bed which is backwashed for cleaning. Many of these systems must be recharged with potassium permanganate which is a very messy, stain-causing material.

Theoretically, this process is supposed to remove 100% of the iron. Practice shows that it removes from 75% to 90% of the iron. Since iron can cause stains at levels as low as 0.3 mg/L, oxidizing filters should be followed by a water softener or KDF® system to remove any remaining iron and hardness that might be present in the water supply.

When iron levels exceed 3 to 6 mg/L, or when colloidal iron or iron bacteria are present, a treatment method known as ozonization

---

IRON

Symbol: Fe

Small amounts healthy, large amounts create a distinctive "metallic-blood-like" taste/odor

Causes reddish stains on plumbing fixtures, laundry

Iron-feeding bacteria create sludge which can damage plumbing

Source: Usually natural

Removal: Aeration, ozonization, catalytic oxidizing filters, ion-exchange, KDF®, chlorine (for bacteria), distillation and reverse-osmosis (for drinking water)

---

or oxidation filtration is required. Iron is "pre-oxidized" by injecting chlorine or ozone into the supply line ahead of a pressure or storage tank. Oxidized iron precipitates in the tank and the resulting "floc" is removed by a filter. If chlorine is used, this must then be followed by a filter containing activated carbon KDF® filter to remove both the chlorine and its disinfection by-products (DBPs) called tri-halomethanes which are carcinogenic. Water softening equipment is installed last in the line, if required, to remove any remaining iron or hardness.

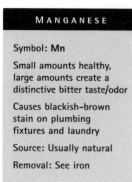

When iron bacteria are detected, chlorination is recommended prior to the installation of water-conditioning equipment. This is accomplished by dumping a half gallon or so of regular household bleach directly into the well and allowing faucets to run until a bleach or chlorine odor is noticeable throughout the system. After the system has been left idle for at least two hours (overnight is preferable), the water is allowed to run until the bleach odor is gone.

If the well is severely contaminated, it maybe necessary to repeat this procedure after 2 or 3 days, then periodically throughout the year. Since most iron bacteria are aerobic, this shot-chlorination procedure controls the bacteria and may eliminate the need for continuous chlorination, especially where low levels of iron are present.

Manganese is frequently found associated with iron, but is rarer than iron. If you remove the lid on your toilet tank and see a red line at the high water mark, the problem is probably iron. If it is black or blackish brown, the problem probably includes manganese. Manganese causes a bitter taste in water, and at concentrations higher than 0.1 mg/L, causes objectionable dark brown or black stains on plumbing fixtures and laundry. The EPA MCL for manganese in water is 0.05 mg/L. The same equipment that removes iron will remove manganese contamination as well.

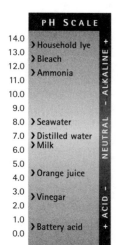

PH SCALE

14.0	> Household lye
13.0	> Bleach
12.0	> Ammonia
11.0	
10.0	
9.0	
8.0	> Seawater
7.0	> Distilled water
6.0	> Milk
5.0	
4.0	> Orange juice
3.0	> Vinegar
2.0	
1.0	> Battery acid
0.0	

ALKALINE + NEUTRAL – ACID +

P H   pH is not a contaminant, but rather an important measure of the acid-alkaline nature of water. A pH of 7 is neutral. Below 7 means the water is acidic, and above 7 means the water is alkaline. Most natural waters range from a pH of 4 to a pH of 9, but commonly measure above pH 7 because of the presence of carbonates and bicarbonates in the water supply.

Understanding the pH of your water is helpful because pH levels of less than 6.5 means that your water is corrosive, and may release metals such as lead, cadmium, and excessive copper from pipes and plumbing fixtures. At levels less than 4, the water is very corrosive and may have a sour taste. At levels exceeding 8.5, water may have an alkali or salty taste, scale may start to form in pipes and equipment, and the formation of trihalomethanes from chlorinated water is accelerated, while the germicidal activity of chlorine is reduced.

SULFATE   Widely found in natural waters, sulfate may be at high levels in areas subject to pollution from mine drainage. Sulfates can be difficult to remove from water and may produce a detectable taste at concentrations of 300 to 400 mg/L, and a medicinal or bitter taste if the concentrations exceed 500 mg/L. At levels exceeding 600 mg/L, sulfates in water act as laxatives, although a tolerance to them can be achieved over time.

SULFATE
Symbol: salt of H_2SO_4
Medicinal, bitter taste
Laxative at high concentrations
Removal can be difficult
Source: Natural, mining
Removal: Distillation, reverse-osmosis, KDF®, catalytic activated carbon

TURBIDITY   Turbidity is caused by the presence of finely divided solid particles and makes water appear cloudy or murky. These particles may be inorganic mineral matter which does not dissolve, or organic matter that has been picked up as the water flows over and through the ground. Turbidity can cause staining of plumbing fixtures and laundry, can have an abrasive effect on plumbing systems, and can interfere with disinfection methods as well as water testing methods. Turbidity is common where the source of water is surface water from lakes, streams or ponds, especially in the spring when runoff from winter snowmelt begins.

TURBIDITY
Cloudy or murky water
Abrasive, stains plumbing fixtures and laundry
Hinders disinfection and water testing methods
Source: Surface water
Removal: Distillation, reverse-osmosis, fine filtration methods

MAJOR TOXIC CONTAMINANTS

Except for certain biological contaminants which can be very toxic even in low concentrations, most toxic substances found in drinking water will produce no ill health effects when consumed in moderate concentrations for short periods of time. However, because many of these toxins (especially heavy metals like arsenic, mercury and lead) accumulate in the body, over time their long term effects may be highly toxic, mutagenic, or carcinogenic, even at low levels.

TOXICITY RATINGS
Long term toxicity rating:
Low:
Moderate:
High:
Extreme:

ARSENIC   Arsenic is a well-known poison that occurs naturally in many rocks, minerals, and soils. Several industrial processes require its use, but most man-made pollution comes from the application of arsenic-containing pesticides. The presence of arsenic in water is not solely the result of pollution. Naturally-occurring arsenic compounds are prevalent in the waters of the Southwest. Areas of New Mexico routinely report arsenic in the water in excess of the MCL of 0.05 mg/L.

The toxicity of arsenic is well known. Like lead, even very small amounts build up in the body over time, causing a condition known as chronic arsenosis. It may take many years for the poisoning to become apparent. Chronic poisoning is usually first noticeable as weakness, tiredness, dry scaly skin, keratosis (changes in skin pigmentation), ganglion cysts, and swelling of the lining of the mouth. Degeneration of the nerves then follows, which produces tingling, then numbing in the hands and feet. Arsenic is also known to cause cancer, and affects the liver and the heart. Chronic arsenosis, in its most extreme form, causes death.

New studies indicate that arsenic in drinking water poses a major risk of cancer, even to people who drink water contaminated with relatively low levels of this widely-occurring element. California state experts found that water containing arsenic at the level of the current EPA standard presents a risk of more than one cancer in every hundred people exposed—10,000 times higher risk than the EPA's standard "acceptable" risk of one cancer in one million people. Tens of millions of people, many of them living in the West and Southwest, drink water every day from their community water systems that contain arsenic at a level of over 2 ppb (parts per billion)—which presents a very significant cancer risk (about one cancer for every thousand people exposed).

Arsenic is effectively removed by distillation, reverse-osmosis, nanofiltration, deionization, certain types of ion-exchange, KDF®, and catalytic activated carbon filtration systems.

BIOLOGICAL CONTAMINANTS All natural water supplies, regardless of source, are likely to contain some microbial organisms. Luckily, only a very small percentage are known to cause disease. These are known as pathogenic organisms. Others, like algae, fungi, and molds, impart foul tastes, odors, or turbidity to the water. Still others are very beneficial, digesting organic debris and removing tastes and odors.

While the use of chlorination to disinfect water supplies has eliminated the plagues of cholera and typhoid that were common before the turn of the century, other chlorine-resistant pathogenic organisms can be quite dangerous when they infect water supplies. these include a wide variety of pathogenic viruses, bacteria, protozoans, amoebas, and parasitic worms. Outbreaks of waterborne disease caused by these organisms are tracked by the EPA, and have been on a steady increase since the late 1960s.

The Centers for Disease Control, in a recent report, found that over 900,000 people become sick in the United States each year from water contaminated with biological organisms. The 1993 Milwaukee disease outbreak alone reportedly affected over 370,000 people; yet the EPA, to date, does not require monitoring for the organism (*Cryptosporidium*) which caused that outbreak. Cryptosporidium is a protozoan which continues to show up in an increasing number of water supplies across the country. This chlorine-resistant pest can overwhelm the body's immune system. As the Milwaukee outbreak and many other cases highlight, AIDS and cancer patients, the elderly, infants, and people who are ill or have compromised immune systems are most at risk. For them, drinking contaminated water can result in death.

Giardia lamblia is another cyst-forming protozoan responsible for the disease known as giardiasis, the number-one waterborne disease in the United States. Giardiasis produces acute diarrhea, sometimes lasting for months, causing so much distress to the system that children, elderly people, and people in weakened conditions can die.

Like Cryptosporidium, Giardia has the ability to create a hard, protective coating known as an oocyst. These cysts have the ability to protect the organism against normal chlorination procedures. Giardia therefore shows up periodically in municipal water supplies around the country.

BIOLOGICAL CONTAMINANTS
Wide variety of pathogenic viruses, bacteria, protozoans, amoebas, and parasitic worms
Short or long term toxicity rating:
☒ ☒ ☒ ☒
Some chlorine-resistant including the oocysts of Giardia lamblia, Cryptosporidium, and bloodworms
Most dangerous to infants, the elderly, and those with compromised immune systems
Source: Fecal pollution, standing water in pipes, unprotected carbon filtration and reverse-osmosis units
Removal: Distillation, reverse-osmosis, nanofiltration, fine filtration with ceramic or carbon block (units that meet NSF Standard 53 will remove worms, Giardia and Cryptosporidium oocysts, bacteria, and some viruses)

Fortunately, Giardia, as well as other forms of protozoan cysts, are easily removed by reverse-osmosis, distillation, or fine filtration equipment using ceramic or carbon block media. If a system is capable of removing these organisms, it will offer a filter labeled at least 1 micron "absolute" (.5 micron or smaller is better) and meet the requirements of NSF Standard 53 for "Cyst Removal." (For a list of filters that meets this standard, contact the National Sanitation

Foundation. See "Resources" at the back of this book.) If the system you use or intend to purchase does not meet this standard, do not rely on it to provide safe drinking water.

Bloodworms are a form of parasitic nematode or roundworm that are also chlorine resistant because of an ability to create cysts. Along with certain other types of worms that are able to resist chlorination, they are appearing in municipal water supplies with increasing regularity. Again, the infective forms of all parasitic worms are best removed by fine filtration at the point of use.

> **If a system is capable of removing Giardia, Cryptosporidium, nematodes and other forms of chlorine-resistant pathogenic organisms, it will meet the requirements of NSF Standard 53 for "Cyst Removal." If the system you use or intend to purchase does not meet this standard, do not rely on it to provide safe drinking water.**

Viruses are the smallest of all pathogenic agents, some being as small as a single molecule. In general, they are both more virulent (dangerous) and more resistant to disinfection than bacteria. Those that occur in polluted water include the adenoviruses, Coxsakie and ECHO viruses, reoviruses, polioviruses, and those that cause infectious hepatitis. Viruses are also harder to treat once an infection occurs, because they are not affected by common antibiotics. With viruses, prevention is truly the best medicine. Some viruses can be removed by fine filtration, but distillation or ozonization are the systems of choice. KDF® also kills many, but not all, viruses—as does ultraviolet light if the system is working properly and the water is not turbid.

The most common waterborne diseases caused by bacteria include typhoid, paratyphoid, Asiatic cholera, bacterial dysentery, tularemia, brucellosis, shigellosis, infectious hepatitis, Weil's disease (jaundice), and anthrax. These diseases are transmitted by consuming water that has been infected with fecal material, and can occur as the result of faulty equipment, lack of maintenance, improperly trained operators or unusual water conditions, i.e., when heavy rains cause surface runoff to contaminate drinking water supplies.

Fecal pollution of a water source can be detected by testing for fecal coliform bacteria. While they are not usually pathogenic in themselves, their presence is a strong indication that other pathogenic bacteria are there and that the water needs to be treated before consumption. All private well owners do well to have a fecal coliform bacteria test performed on their wells at least once a year.

CHLORINE   Chlorine is a highly effective and inexpensive disinfecting agent used extensively in the United States to treat municipal and individual water supplies. It has played a key role in eradicating water-borne infectious diseases such as typhoid and cholera. But while chlorination has helped save many lives, a growing body of evidence indicates that the practice may be very hazardous to human health, especially when high levels of free residual chlorine leave the water treatment facility and arrive at your tap.

In 1900, before chlorination was in general use, there were about 35,000 deaths from typhoid in the United States. In that same year, there were over 68,000 deaths from heart disease, or about 137 per 100,000 population. By 1955, the death rate from typhoid had dropped to only 50 people, but those dying from heart disease had skyrocketed to 536,000, or 355 per 100,000 population. Today, over 50% of those who die are victims of heart disease. Last year, that amounted to well over a million people in the United States. In fact, more American men died of heart disease in the last two years than have been killed in every war since the Declaration of Independence was signed.

It is absolutely clear that the presence of free chlorine in water creates heart disease. In fact, the dramatic rise of heart disease in this country directly correlates with the use of chlorination to disinfect water supplies. Since the early 1950s, scientists have known that the presence of free chlorine in water is a primary cause of, and completely linked with, the development of atherosclerosis. It causes fats to form the cholesterol deposits known as plaque. It is this plaque that clogs arteries, resulting in heart attacks and strokes.

Further studies demonstrate that chlorine destroys vitamin E, an essential antioxidant. Vitamin E prevents the formation of free radicals, highly destructive molecules that promote tissue breakdown and tumor formation. These free radicals come from oxidized dietary fats, synthetic chemicals, and from various immune system activities. A lack of vitamin E impedes heart muscle function, weakens capillary walls, and constricts blood vessels, thereby elevating blood pressure and increasing the risk of heart disease.

Scientists from the Oak Ridge National Laboratory in Oak Ridge, Tennessee, reported that a study of 1,520 residents from 46 Wisconsin towns had linked the drinking of chlorinated water to the formation of high cholesterol.[9] Women were shown to be at greater

CHLORINE
Symbol: Cl
Long term toxicity rating: ☠ ☠ ☠
Effective disinfectant
Distinctive taste, odor
Carcinogenic, creates disinfection by-products that are poisonous and carcinogenic
Increases risk of heart disease
Destroys vitamin E
Toxic to friendly intestinal bacteria
Source: Public water systems, pools, baths and showers
Removal: Distillation, reverse-osmosis, ozonization, activated carbon, KDF®, catalytic activated carbon

It is absolutely clear that the presence of free chlorine in water creates heart disease. In fact, the dramatic rise of heart disease in this country directly correlates with the use of chlorination to disinfect water supplies.

risk than men. At the same conference, J. Peter Bercz of the EPA's Health Effects Research Laboratory (HERL) described various "abnormalities" in fat metabolism of mice drinking highly chlorinated water (15 ppm) and eating diets with a fat content comparable to that of the typical American diet. The chlorine caused a noticeable increase in low-density lipo-proteins (bad cholesterol), demonstrating that chlorinated water alters the way the body metabolizes fat, thus increasing the risk of heart disease.

Other evidence linking chlorinated water and health problems is equally disturbing. Studies show that chlorinated water is toxic to human intestinal bacteria which convert organic compounds in our food into necessary nutrients. It may also deplete the small intestine of bacteria which produce vitamin B_{12}.

Disinfection by-products (DBPs) are formed when chlorine is

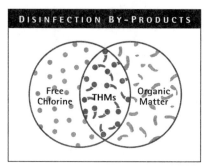

DISINFECTION BY-PRODUCTS

Free Chlorine — THMs — Organic Matter

used to disinfect water. One class of these chemical compounds is called trihalomethanes (THMs), which are formed when chlorine reacts with organic matter in water. A recent study by doctors from Harvard and the University of Wisconsin, published in a prestigious journal, found that DBPs may be responsible for 10,700 or more rectal and bladder cancers per year.[10] Another massive study of pregnant women and their babies, led by doctors from the Public Health Service, found that certain birth defects are significantly associated with DBPs, and urged that follow-up studies be conducted as soon as possible.[11]

Despite the clear evidence that DBPs pose serious risks, the EPA has rules controlling only one class of them (trihalomethanes), and only in large drinking water systems. New rules for DBPs may not be issued until 1996.

THMs (i.e. chloroform and trichloroethylene) are known to damage the kidney, liver, and nervous system, and to cause birth defects. Many THMs are proven to be potent carcinogens (cancer-causing agents). How dangerous are they? One way to look at it is that your chances of being shot by a person with a handgun are much less than your chance of developing cancer from THMs in water.

In addition to THMs, HERL scientists have found that two other potent carcinogenic substances are formed when chlorine interacts with organic material created by the decay of plants. One of these, known as MX, has shown up in every chlorinated drinking water source tested for it. According to EPA chemist H. Paul Ringhand, MX may be the single largest mutagenic (the ability to induce genetic mutations—a rough gauge of cancer-causing potential) agent in municipal water supplies. Another mutagen, called DCA, is known to alter cholesterol metabolism and cause liver cancers. Many other scientists are also convinced that these two chlorinated acids are the most dangerous chemicals one can be exposed to. Both are found primarily in chlorinated drinking water.

> Many scientists are convinced that the chlorinated acids MX and DCA are the most dangerous chemicals one can be exposed to. Both are found primarily in chlorinated water.

In 1987, the Journal of the National Cancer Institute reported a study showing that the drinking of chlorinated water increases one's risk of developing bladder cancer by as much as 80%; other studies link it to the development of colon cancer as well. More recent studies also suggest that chlorine in water supplies maybe responsible for as many as a third of breast cancers in women.

Not only is drinking chlorine dangerous, but hot showers and baths can release 50% of dissolved chlorine and up to 80% of THMs like chloroform and tetrachlorethylene, which are then breathed into the lungs or absorbed into the body through the skin. This absorption through the skin and lungs via hot showers and baths is a primary cause of poisoning due to toxic materials in water. Special shower filters containing KDF® are an inexpensive way to solve the problem. As with all toxic exposure, small children, elderly people, and those with weakened immune systems are most at risk.

> Not only is drinking chlorinated water dangerous, but hot showers and baths can release 50% of dissolved chlorine and up to 80% of THMs, which are then breathed into the lungs or absorbed into the body through the skin.

Because of their experience in World War I with chlorine gas used as a chemical weapon, communities in Europe use ozone and ultraviolet light to disinfect water supplies. Holland uses sand filtration to deliver highly purified water to its citizens. Clearly, other methods of disinfection of water supplies are available; they just cost more than most municipalities are willing to spend.

In spite of promises made by politicians, it is likely to be decades, if ever, before non-chlorinated water is routinely available from the tap. In fact, the current political climate is actually working to weaken existing pollution laws. Until such time as the government is forced into upgrading water delivery systems, common sense dictates that individuals protect themselves with the purchase of bottled drinking water or point-of-use filtration systems that are effective in the removal of chlorine and its associated DBPs, which should be removed not only for drinking purposes, but for bathing purposes as well.

Chlorine, and some but not all of its DBPs like THMs or chloramines, can be removed by granular-activated carbon. Better yet is KDF®, which can remove many times more chlorine, as well as all DBPs. While high quality reverse-osmosis and distillation-based systems remove chlorine and DBPs, home units are impractical for treating water used for bathing. See section entitled "Understanding Your Choices," for more information.

FLUORIDE   New studies pointing to potential drawbacks in water fluoridation have rekindled a debate that most scientists want to see closed. Many new studies (as documented below) point to the significant danger of ingesting fluoride. One study has prompted the FDA to warn parents not to let children ingest fluoride in toothpaste while brushing teeth. Another demonstrates that people who use fluoridated water in aluminum cookware release up to 1,000 times more aluminum into their food.

Opposition to fluoridation has always been treated as an anomaly, to be explained away as the activity of fringe groups and quacks. Nevertheless, not all critics of fluoridation can be explained away as quacks, nor have they been effectively silenced.

Brian Dementi, toxicologist at the Virginia State Department of Health, in his 1981 report *Fluoride in Drinking Water,* discussed the many scientific papers showing fluoride to be both mutagenic and carcinogenic. Albert Burgstahler, a Kansas biochemist and co-author, with the late George Waldbott, M.D., of *Fluoridation: The Great Dilemma,* reinforces that view by demonstrating that pathological changes in the kidney can be produced by fluoride levels of less than 100 ppm.

One of the largest epidemiological studies ever done on fluoridated water and carcinogenicity was conducted in 1977 by Dr. John Yiamouyiannis and Dr. Dean Burk. Dr. Yiamouyiannis was the former biochemical editor of the prestigious journal of the Chemical Abstracts Service, the world's largest chemical information center, but was removed when federal funding was jeopardized because of his propensity to publish information critical of fluoridation. Dr. Burk was the retired head of cytochemistry at the National Cancer Institute. In their study, Drs. Burk and Yiamouyiannis monitored

---

FLUORIDE

Symbol: Ionized fluorine, F

Long term toxicity rating:
☒☒☒

Mutagenic, carcinogenic; can cause "brittle bone" syndrome and mottling of teeth

Destroys enzyme function

Source: Public water systems, toothpaste, nature

Removal: Distillation, reverse-osmosis, deionization, activated carbon, KDF®, catalytic activated carbon

cancer rates over a twenty-year period in ten fluoridated American cities and ten non-fluoridated. What they found was an increased mortality rate from cancer in those living in cities with fluoridated water. New studies indicate that fluoride levels of as little as 1 ppm can cause chromosomal damage by inhibiting or interfering with the ability of DNA to repair itself.

Does fluoride prevent tooth decay? Perhaps, but contrary to popular opinion, the jury is still out. According to the World Health Organization, the rate of reduction in tooth decay is greater in Finland and Norway where water is not fluoridated than in the U. S. where water is routinely fluoridated. A massive two-year, multi-million dollar, taxpayer-funded study[12] of nearly 40,000 children in all areas of the country was conducted during the late 1980s by the National Institute of Dental Research (a department of the United States Public Health Service) in order to silence critics of fluoridation once and for all. Instead, what this extensive, peer-reviewed study found was that fluoride has no effect on the incidence of tooth decay! The results of this study were buried. In fact, a great deal of solid, peer-reviewed information critical of fluoridation has been buried.

As reported in *Newsweek* (Feb. 5, 1990), health information critical of the use of fluoride in drinking water has been intentionally withheld from the public, including many dental professionals. Only recently has the information, available since 1985, been pried loose from the U. S. Public Health Service, through petitions based on the Freedom of Information Act. It seems that those who are so anxious to put their reputations on the line in defense of fluoridation are not so anxious to inform the public of studies confirming its dangers. Considering the potential legal ramifications of intentionally adding what may prove to be a serious poison to the drinking water of 55% of American households, perhaps we can understand why.

> As reported in Newsweek (Feb. 5, 1990), health information critical of the use of fluoride in drinking water has been intentionally withheld from the public, including many dental professionals.

In any event, the average consumer of fluoridated water would do well to become aware that sodium fluoride, or hydrofluosilicic acid, is rated as more toxic than lead in chemistry indexes, and only slightly less toxic than arsenic. Like chlorine, it is a halogen that destroys enzyme function. It is not an essential nutrient, and has never been shown as necessary for human life. It also wouldn't hurt to know that sodium fluoride is a by-product of aluminum manufacture, and that the transformation of sodium fluoride from dangerous chemical to benign cavity fighter came as a result of promotion from the Mellon Institute, the chief research facility of ALCOA Aluminum Company, North America's largest fluoride producer.

Many countries in Europe have either banned fluoride or discontinued its use. They see the debate over whether fluoride is toxic at 1 or 2 ppm as missing the point. To them, the increased risks of cancer and skeletal fluorosis known to be caused by fluoride far outweigh any potential benefit to one's teeth. Their approach is to implement sound nutritional principles, remove pollutants that compromise natural immunity, and fund appropriate educational programs designed to resolve not only problems with tooth decay, but many other modern degenerative diseases.

Fluoride can be removed from your water by carbon filtration, reverse-osmosis, distillation, or a system employing KDF®.

HYDROGEN SULFIDE

Symbol: H_2S

Long term toxicity rating:
☠ ☠

Strong "rotten egg" odor

Poisonous and flammable in high concentrations

Corrosive to plumbing

Source: Decaying underground organic material

Removal: Distillation, reverse-osmosis, aeration, ozonization, catalytic activated carbon filtration

HYDROGEN SULFIDE Hydrogen sulfide is a gas with a rotten egg odor. In the Southwest, it often occurs in tandem with iron and manganese. It is very corrosive to plumbing, kills aquarium fish, tarnishes silver, and at high levels is dangerous to health. It can be tasted at levels as low as 0.5 ppm, and in concentrations that are dangerous, its odor will be quite obvious. When present it must be removed before the water is suitable for human use. This is usually accomplished with reverse-osmosis, aeration, ozonization and chlorination equipment, or catalytic activated carbon filtration units. The EPA maximum contaminant level for this substance is 0.05 mg/L.

LEAD Lead is one of the most dangerous pollutants found in drinking water. It is dangerous not only because of its effects, but because of the widespread nature of its distribution. Healthcare researchers estimate that as many as 20% of Americans are exposed to dangerously high levels of lead in their drinking water. Overall, the EPA has found that controlling lead-contaminated drinking water could reduce lead exposure for between 130 and 190 million Americans.

Lead is responsible for kidney, brain, and central nervous system disorders. In adults it can cause miscarriages, hypertension, multiple sclerosis, impotency, numerous nervous system disorders and even death. These and other health effects of lead at low levels have been well-documented; therefore, the EPA established a goal of zero lead in drinking water because "there are no clearly discernible thresholds for some of the non-cancer health effects associated with lead."

Young children, especially bottle-fed infants, and fetuses are particularly vulnerable. The EPA found that more than 85% of the lead found in the blood of bottle-fed infants may be derived from baby formula made with lead-contaminated water. For infants, consuming even very small amounts of lead can lead to irreversible brain damage, intellectual, emotional, and developmental problems, numerous nervous system impairments, and stunted growth. High levels of lead have been found in children suffering from lethargy, personality aberrations, and mental retardation.

Lead gives a sweet taste to water, causing people to mistakenly believe that their water is particularly "good." Since lead accumulates in the human body, particular care must be taken to ensure that lead does not enter your water supply.

Lead leaches into drinking water in a variety of ways. Many older communities have lead supply lines to homes. Over 90% of all U.S. homes have lead in their plumbing in some form, either as lead service pipes, lead-soldered pipe connections, or lead in the brass alloys of which faucets, well pumps and other plumbing parts are made. Although a federal ban against the use of lead in plumbing systems took effect in June of 1988, it is frequently ignored by plumbers who continue to use lead solder, and manufacturers of water delivery equipment (like well pumps and water fountains), who use metal alloys containing lead. As recently as April 18, 1994, the EPA issued a warning to hundreds of thousands of people who drink from private wells to switch to bottled water and test for lead contamination from submersible pumps containing these alloys.

The maximum amount of lead permitted by the EPA is 0.05 mg/L. In fact, according to the EPA, no amount of lead in drinking water can be considered "safe." Municipalities do test for lead, but they draw their samples at the source, and lead is a contaminant that is most likely to show up after it has left the treatment plant, passed through city piping, your home piping, and in your glass after passing through your faucet.

Until you are certain that your water does not contain lead, switch to bottled drinking water or let water run for a minute or two to

---

LEAD

Symbol: Pb

Long term toxicity rating:

☒ ☒ ☒ ☒

Poisonous, especially dangerous to infants and the elderly

Contributes to many serious health disorders

Accumulates in tissues

Imparts sweet taste to water

Source: Lead in pipes, solder in plumbing, brass in faucets, fixtures, well pumps

Removal: Distillation, reverse-osmosis, KDF®, ion-exchange resins, catalytic activated carbon briquette or block media make sure that any dissolved lead that may have accumulated in your home's pipes (known as "first draw" water) has been flushed away before drinking.

Have your water supply tested; if lead is present, install a filtration system capable of removing it from primary sources of water used for drinking and cooking purposes (usually the kitchen sink and home bars). Make sure your water treatment system has a lead-free faucet. Believe it or not, most residential systems dispense water through spigots made of lead alloys which leach lead into the highly purified (and therefore aggressive) water they dispense.

It is also not a good idea to regularly drink water from water fountains or other tap water sources in schools and government and commercial buildings, especially if they were built prior to 1988, unless you know they have been tested and are certified to be free of lead contamination. If you have schoolchildren, it is worth demanding to see written proof that their school's water is free of lead—many aren't.

NITRATES/NITRITES   Nitrate nitrogen is found in many water supplies and is especially harmful to infants and elderly adults. In the first three months of life, an infant's digestive system is populated with bacteria which convert nitrates to nitrites. Nitrites bind strongly with blood hemoglobin (the part of red blood cells which transports oxygen to the body's cells) and prevent sufficient oxygen transport in the baby. Shortness of breath, susceptibility to illness, heart attack, or even death by asphyxiation can result. By six months of age, hydrochloric acid concentrations in the stomach rise to the point where the nitrate-reducing bacteria are killed. Nitrates are therefore not as much a concern in older children and adults. Later in life stomach acid production decreases. That's when nitrate-reducing bacteria can repopulate the digestive tract and why elderly people and those with low stomach acid levels are also at risk.

Nitrate concentrations as low as the EPA limit of 10 mg/L can cause the death of infants and susceptible adults. The condition is called methemoglobinemia (aka: blue baby syndrome). Besides shortness of breath, another common symptom of the illness is blueness of the skin called cyanosis.

Nitrate nitrogen enters the water supply by seeping through soil containing nitrate-bearing minerals, fertilizers, plant debris, or the products of bacterial decomposition. It is particularly prevalent in areas downstream from farming communities. Its presence in other areas can mean that the water source is contaminated with human and/or animal wastes, such as effluent from septic systems or cesspools. Surveys conducted by the EPA indicate that nearly one million households on private wells periodically use water that exceeds the EPA nitrate standard of 10 parts per million (ppm).

Whether ingested from water or foods like bacon and lunch meat, nitrites combine with stomach acids to create a class of compounds known as nitrosamines. Some nitrosamines have been shown to be potent tumor producers in rats, but no studies have been done to determine the effect in humans.

While adults can tolerate more nitrates/nitrites, elderly people, people with respiratory dysfunction of any kind, and families with children do well to avoid all nitrates/nitrites in their water.

Nitrates or nitrites may be removed by the processes of distillation, deionization, reverse-osmosis, or strong-base anion resin exchange. Anion resin systems are regenerated much like water softeners and are designed for whole-house treatment. For best results, water softeners may also be necessary to prevent fouling of the anion resin system from the precipitation of calcium and magnesium salts.

NITRATES/NITRITES

Symbol: NO_3, NO_2—ions from nitrogen salts

Long term toxicity rating:

May cause death in infants

Source: Fertilizers, septic systems, animal waste

Removal: Distillation, deionization, reverse-osmosis, or strong-base anion resin exchange

R A D O N   Overall, radon poses a greater health risk than any other environmental pollutant. Radon is a naturally-occurring radioactive gas which is a product of the decomposition of uranium, a widely dispersed element in the earth's crust. The EPA estimates that waterborne radon may cause more cancer deaths than all other drinking water contaminants combined, and that at least 8 million people may have undesirably high radon levels in their water supply.

There is a lot of disagreement about how much radon in water is dangerous to human health. A widely-used rule of thumb is to definitely take action if the level in your water exceeds 10,000 picocuries/liter. That level is roughly equivalent to 1 picocurie in indoor air. Because you breathe in more radon during one shower than in all the water you could drink in a week, to effectively remove radon you must treat all the water in the house.

Radon levels are best controlled by ozonization and aeration units that treat all the water in the house and vent the dangerous gas to the outdoors.

RADON

Symbol: Rn

Long term toxicity rating:

Radioactive gas

Highly carcinogenic

Source: Natural, uranium mining

Removal: Aeration, ozonization

8

While water quality is a serious concern, it is a problem that every responsible person can easily solve for themselves. Modern technology has placed excellent water conditioning systems within the economic reach of nearly every American.

Understanding
Your Choices

LUCKILY, TODAY'S HOME WATER treatment devices are advanced enough to render harmless even the foulest domestic supply, but the more complex the contamination, the more expensive the solution. At the time of this writing, initial investment in quality equipment is likely to start at around $400, moving to over $6,000 for seriously contaminated water. Once installed, many systems have to be constantly monitored and maintained.

Be aware that, according to the EPA, the word "purifier," when applied to water treatment devices, implies a product proven effective against bacteria, viruses, and cysts. Water purifiers must be registered with the Antimicrobial Program Branch of the EPA's Office of Pesticide Programs. As a result, the majority of systems will not refer to themselves as "water purification systems," but as "water treatment systems" or "water conditioning systems" or some other such name. No matter what they call themselves, in order to assure yourself of healthy, pure water, read and understand the following information before shopping for a system. Then, when shopping, use the checklist provided at the end of this section to be sure you are getting what you need at the best possible price.

BOTTLED WATER While bottled water is not a "system," it has become the alternative of choice for a growing percentage of the population. As problems with water have invaded the public consciousness, sales of mineral waters, flavored waters, sparkling waters, as well as normal water, has been growing at two to three times the rate of the rest of the beverage industry. In 1990, Americans spent $2.2 billion for more than two billion gallons of bottled water.

Some people buy bottled water because they object to the taste, smell, color, or odor of tap water. Others are confused by the alternatives, or have no desire to permanently install expensive equipment in rented or otherwise temporary accommodations. New technology has rendered this latter problem obsolete, but many consumers remain uneducated to the alternatives.

Most bottled water is regulated by the federal Food and Drug Administration (FDA). In early 1993, that agency proposed stricter rules for labeling different types of bottled water—artesian, distilled, spring, mineral, and purified, for example—and required that they meet all federal drinking water standards. These new rules are a response, in part, to the 1990 discovery of benzene (a highly toxic chemical) in Perrier, a popular and premium brand-name bottled water.

Mineral waters are exempt from the drinking water standards because their main selling point—high levels of calcium and other minerals—would put them in violation of the standards. The new rules create a standard definition for mineral water, requiring it to come from a protected (not identified) source and contain at least 250 ppm of TDS. Significant amounts of nutrients like calcium, sodium, or iron are required to be printed on the label.

Except for certain states like California, Pennsylvania, and Florida, that have adopted stricter regulations, bottlers selling products in only one state are regulated by state health authorities, not the FDA. Bottlers belonging to the International Bottled Water Association also undergo unannounced inspections by an independent laboratory. No other regulations exist for bottled water companies, except the same standards that apply to any purveyor of public water. These standards are mandated by the United States Safe Drinking Water Act, which specifies that public water supplies must contain less than the maximum contaminant levels (MCLs) for certain substances.

> By law, bottled water need be no cleaner than regular tap water; in fact, in spite of all the talk about "mountain spring water" etc., depending upon the region, 25% to 75% of bottled water is nothing but tap water bottled in one city and shipped to another.

By far, the biggest health concern with bottled water is the lack of appropriate sanitation. The problem of sanitation is of special concern among those firms that bottle reverse-osmosis, distilled, or deionized water, whether operating as a customer fill station for customers with their own bottles, or whether the product is bottled for sale in stores. That's because unchlorinated water provides an ideal breeding ground for pathogenic bacteria, which multiply rapidly (populations can double every 20 minutes or so) in the oxygen-free environment provided by bottling. Over time, pure water (free of dissolved minerals) is an aggressive solvent, so it will leach carcinogenic chemicals out of the typical opaque plastic containers in which it is stored, especially when the container is exposed to sunlight. According to a pilot study done by the EPA on 25 randomly-selected bottling establishments making approximately 50 bottled water products, serious sanitary deficiencies were found at every single bottling plant visited. They found that plastic bottles, shipped without caps, arrive at the plant in cardboard cartons. These bottles are stored and handled in the open air, become contaminated with airborne contaminants like dust mites and bacteria, and are filled without rinsing. Caps are often placed by hand, directly from the cartons in which they were shipped, without any procedure used to disinfect either the bottle, cap, or the hands of the person placing the cap. Employees are not tested for disease, nor are they required to avoid the bottling area when they are sneezing from colds or have open cuts or infections on their hands.

A random study conducted by the EPA found that employees of bottled water companies are not tested for disease, nor are they required to avoid the bottling area when they are sneezing from colds or have open cuts or infections on their hands.

At every facility tested, the lack of employee sanitation and facility maintenance resulted in product water that was subject to contamination, not only from the containers, but also from the physical surroundings and the people who come in contact with the bottling operation. The result of tests on the subsequently bottled water was predictable. About one in ten of the samples showed the presence of pathogenic bacteria. When the water was stored for 63 days and then tested, *every* sample showed the presence of large numbers of pathogenic bacteria.

Further tests revealed that the labeling of bottled water did not correspond with the contents revealed by chemical analysis. Those analyses revealed significant amounts of one or more of the following in every sample of water tested: chlorine, sulfate, nitrate, copper, manganese, lead, iron, zinc, mercury, and arsenic.

Other concerns relate more to convenience and expense than health. Bottled water is expensive, hard to carry around, and bulky to store. If you want to minimize expense, you can buy 1, 2.5, and 5-gallon containers and fill them yourself at water dispensing machines in health food stores and elsewhere. However, full 5-gallon containers weigh in at about 45 pounds, making them a challenge for any but the strongest people to carry around. Those who are incapacitated, elderly, or ill (and therefore need clean water the most), are least likely to be able to deal with that sort of problem. Delivered, or purchased in the store in 1-gallon containers, the price of bottled water zooms upward to well over $1.00 a gallon, making it unaffordable to many.

If you purchase containers for water, either use glass (can be dangerous if dropped) or clear plastic, which does not leach as many chemicals into the water as opaque plastic containers. Store out of sunlight for no longer than a few days.

If you intend to store water, either can it using standard methods for canning things like vegetables, or chlorinate it. Unless it has been canned, do not drink water that has been stored for longer than 60 days without filtering it, using carbon block or ceramic media certified to remove all contaminants over .5 microns, or boiling it for twenty minutes to kill pathogenic bacteria.

In general, considering its lack of reliability, its expense and inconvenience, bottled water is better than nothing, but for most people, it is not a satisfactory solution.

CARBON FILTERS   Filtering water through a medium of carbon is one of the oldest and most tested of all water filtering technologies. Centuries ago, sailors noticed that their drinking water supplies stayed fresher and tasted better when stored in charred barrels. The effectiveness of carbon filters depends on the kind of carbon used, types and amounts of contaminants, and water temperature.

Activated carbon is often mistakenly referred to as charcoal. It is not. It is a charcoal-like substance made from coal, wood, or even petroleum products specially treated with steam and high heat in the absence of oxygen. The resulting material is extremely porous and has a remarkable surface area (125 acres to a pound). Contaminants are not absorbed by the carbon, but adsorbed, which means they cling to its surface. Generally, activated carbon has the capacity to remove volatile organic chemicals (VOCs) like some pesticides and industrial chemicals, and most halogenated organic compounds like PCBs and PBBs. It is effective at removing chlorine, but not compounds created by chlorine (trihalomethanes like chloroform). Some forms will remove some, but not all, heavy metal contamination. Activated carbon will not remove most biological contaminants, asbestos particles, fluorides, nitrates or other salts, nor many inorganic chemicals.

Technology has improved on activated carbon with a new product referred to as catalytic/adsorptive carbons. These carbons are manufactured using a patented process that modifies the electronic properties of the carbon surface, while maintaining its adsorptive pore structure. The result is a product that not only adsorbs contaminants, but provides an environment where oxidation/reduction reactions (exchanging of electrons) can act to promote a range of chemical reactions where conventional carbons are ineffective. Unlike conventional carbons, catalytic/adsorptive carbons are effective at oxidizing hydrogen sulfide gas (H_2S), peroxides, chloramines, and hydrazine. In these applications and others, catalytic/adsorptive carbon can reduce the time water must be in contact with the media, extend media life, and reduce the amount of media required, thereby decreasing equipment size.

Perhaps the biggest drawback to activated carbon or catalytic/-adsorptive carbon filters, used on their own, is that if the water contains bacteria, they tend to enjoy the environment that the carbon filter provides, and will contaminate the system by growing in the filter. Some carbon adsorption units utilize a silver-nitrate-impregnated activated carbon to presumably kill bacteria. Silver nitrate is a potential toxin in its own right, so units using this technology have to be registered with the EPA as having been tested for the leaching of unacceptable amounts of the substance into the drinking water.

Carbon filters may use granular media, or the carbon may be compressed into blocks. To understand the difference, visualize water slowly seeping through a solid block of wood, as opposed to running through sawdust. Because water has more contact time with carbon in the block form, it is generally the most effective type of carbon filter. The best employ media that filter all particles down to .4 micron in size, thus making them suitable for filtering dangerous biological organisms like Cryptosporidium and Giardia oocysts.

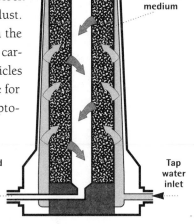

Carbon block filtration medium

Filtered water outlet

Tap water inlet

Carbon Block Filter

For any form of carbon filter to work well, all the water flowing through the filter must have contact time with the media long enough for impurities to be removed. Some of the cheapest units simply run water straight from the tap through a bed of carbon and into your glass. In this type of filter, water is free to form "channels," or stream beds, through which it can flow without ever contacting the media. This type of filter may remove some odors and tastes, but otherwise accomplishes very little. What's worse, many of these filters are sold at prices three to five times higher than necessary, by people working for multi-level marketing firms who are often poorly educated in water treatment issues.

Over time, carbon's ability to remove impurities gradually degrades, because the pore spaces become saturated with impurities. If the filter stopped functioning when it became inefficient, it wouldn't be such a problem. In fact, however, you can't tell when the carbon is exhausted. As a result, most people who have carbon-based units don't replace the filter cartridges frequently enough. Not only are they not getting the benefit they expect, they may actually be introducing harmful bacteria into their water.

Because they employ so little media, systems which use filters which fit on top of your faucet or on top of a glass carafe are virtually useless at doing anything but improving the taste and odor of water, and are potentially hazardous. A better choice is a unit which uses standard 10-inch cartridges. These help remove chlorine and other taste and odor causing contaminants for about six months for the average family using municipal tap water. By themselves, however, they do little or nothing to remove harmful biological agents, disinfection by-products and heavy metals, and may actually breed harmful bacteria. In any case, cartridges must be replaced frequently to be of real value.

> **Most people who have carbon-based units don't replace the filter cartridges frequently enough. Not only are they not getting the benefit they expect, they may actually be introducing harmful bacteria into their water.**

Carbon is best used as a filtering agent in combination with some other technology that can compensate for its weaknesses. KDF® is a media that is ideal for use in combination with carbon. Placed ahead of carbon in a system, it effectively removes chlorine and DBPs like trihalomethanes and chloramines, heavy metals, fluoride, and many compounds and chemicals that carbon can't touch. This leaves the carbon free to do its best on VOCs like pesticides, herbicides, PCBs and PBBs, and remove substances that affect the taste and odor of water. KDF® ahead of the carbon also renders the water bacteriostatic, unable to support the growth of bacteria and other biological contaminants.

Many water softeners include activated carbon in their resin beds to remove chlorine and enhance the performance of the system. Unfortunately, the carbon in such units becomes depleted in a couple of years, while the resin material continues to function for a decade or more. As a result, in the first couple of years after installation, the performance of even the most expensive water softeners declines substantially in the ability to remove chlorine and organic pollutants from water.

In the best systems, a .5 micron absolute carbon block cartridge is combined with KDF® redox media to create a dependable supply of clean water. The carbon block is used as a post filter to remove organic materials and biological contaminants such as Crypto-sporidium oocysts not affected by KDF®, while the KDF® effectively removes chlorine, heavy metals, and other contaminants that tend to clog up the carbon. Best of all, KDF® renders the water bacteriostatic, reducing the worry that the carbon will become a breeding ground for harmful organisms. Used in this manner, the carbon may last from three to five times longer than without KDF®. Simple carbon units are quite inexpensive to install and maintain. Many systems are available for under $100, and can be refilled for as little as $20 per filter replacement. Similar units sold by multi-level marketing firms may cost nearly $200 and must be thrown away after they become depleted, sometimes in as little as three to six months. Avoid them.

CERAMIC FILTERS   These units use a special porous ceramic material to provide a durable, ultrafine, filtration medium. They work well to remove bacteria, cysts, asbestos fibers, algae and sediment from water. Unlike many other filtration media, when they become clogged they can be brushed clean and used several times. They do little or nothing to remove chlorine, chloramines, or other disinfection by-products from water. Neither do they remove nitrates, hydrogen sulfide or heavy metals. Since these types of filters are used primarily in portable units designed to purify water for backpackers and campers, the removal of chlorine is not important, but the consumption of nitrates, hydrogen sulfide, volatile organic compounds and heavy metals should be avoided, especially over long periods of time. Because of their limitations, ceramic filters are not generally used to purify household water without some other form of filter that can remove what ceramics can't.

DISTILLATION   Distillation is the age-old process that extracts gasoline from crude oil and moonshine from corn mash. A heater heats water to its boiling point in a boiling chamber, sending steam into a condensing device where it cools back to its liquid water form. The problem is that any substance with a boiling point lower than water will co-distill with water and re-contaminate it at even higher concentrations than before. Unfortunately, many volatile organic chemicals (VOCs), trihalomethanes (like chloroform), and numerous other contaminants all have lower boiling points than water. That makes distillation a good choice for removing salts and minerals, but a poor one for removing organic chemicals.

For water to become truly pure, distillation must be preceded by carbon or KDF® filtration, or employ a fractional distillation method (venting of other lower-boiling-point compounds prior to extraction of water vapor). If you are going to buy a distiller, make sure the one you are considering offers one of these methods for extracting volatile organics, because many don't. Another significant disadvantage is that the process is very energy-intensive, and therefore expensive, relative to other methods of water treatment. Also, because this technique leaves behind the inorganic salts, the apparatus often becomes coated with mineral scale, requiring frequent and difficult cleaning. Finally, since a distiller generates a lot of heat, it can be less than welcome during warm weather.

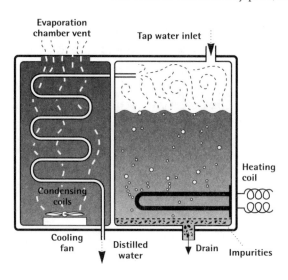

Schematic of a Distillation System

The best distillers produce the purest drinking water available, but remove healthful minerals and cause the water to taste "flat." Relative to other methods, distillers are also expensive to buy, maintain, and operate, and some produce a less-than-satisfactory product.

ION–EXCHANGE SYSTEMS (Water Softeners) This technology uses a bed of electrically charged resin designed to hold sodium ions in suspension. As water finds its way through the resin bed, calcium and magnesium ions in the water are replaced by sodium ions. (Potassium salts may also be used instead of sodium chloride.) As this process occurs, the resin bed eventually becomes saturated with contaminants and must be "recharged." This is accomplished by backwashing the resin bed with a heavy brine to release calcium and magnesium ions, and replace them with sodium ions ready for a new round of water. With most units, a timer is set to start this backwashing process at a time when water is not likely to be in demand (around 2 a.m.). Some units split the resin bed into two tanks, one of which backwashes while the other processes water. Either technology works well.

With any ion-exchange system, the more hardness ions the water contains, the more salt it will take to replace them. Some users in the Southwest find it necessary to add 100 or more pounds of salt per month to keep the conditioner properly charged. That costs around $10 to $15 per month, and the 40-pound bags may be a little heavy for some people to handle.

Older model softeners tend to use far more salt than necessary. New generation softeners combine flow meters and timers to recharge only when a predetermined amount of water has been used. Known as "demand initiated regeneration systems," or DIRs, they can save as much as 50% on salt while keeping the amount of salt in finished water to a minimum. Unfortunately, not all companies use the new technology.

In most locations, new generation DIR systems add less salt to the finished water than is contained in mother's milk. For those concerned with adding sodium to their diet, potassium chloride can be substituted for sodium chloride, with equal results. It is more expensive however, usually costing about three times more than normal salt.

In extremely hard water areas, quality water softeners quickly pay for themselves. They not only save money in cleaning products, but they greatly prolong the life of household plumbing, fixtures, and water-using appliances—especially hot water heaters.

Softening (ion-exchange) is not a treatment designed to purify water for drinking. It does not remove dangerous organic chemicals, most heavy metals, or biological organisms that may contaminate water. It is best used to soften water for economic reasons (see *Hardness* in Chapter Seven). For drinking water, the system should be followed by a filtering system designed specifically for water purification purposes. KDF® combined with .5 micron absolute carbon block or reverse-osmosis systems are excellent choices. Because most ion-exchange units do nothing to remove harmful chlorine from the water, KDF® redox units should also be installed on showers and tubs to eliminate absorption of chlorine through the skin and lungs.

Some national chains sell systems that combine a water softener with a reverse-osmosis unit for $3,500 or more, installed. That's totally unnecessary. Much of that money pays for sales commissions and the other costs of their expensive marketing methods—not equipment. As good or better quality systems can be purchased for much less. (See Chapter Nine, "Shopping Tips.") A common gimmick used to justify these companies' high prices is the claim that their system will reduce the cost of cleaning products as much as 90%. To drive home the point, a customer may be offered a several years' supply of soap products if they purchase the system. While soap savings may be substantial in areas that suffer from extreme hardness, and a good case can be made that softeners pay for themselves if hardness exceeds 10 to 12 gpg, it still doesn't mean you need to pay 200% to 300% more for a system than is necessary.

Unless you need to deal with water that is excessively hard (over 10 grains per gallon hardness), there is no need to spend money on a softener/reverse-osmosis combination of equipment to begin with. For as little as $300 to $400, you can buy completely adequate systems employing a combination of KDF® redox media and granular-activated carbon in one cartridge, and a .5 micron absolute carbon block media in another. These systems will remove as much chlorine and DBPs, VOCs, heavy metals, biological contaminants and other problem contaminants as reverse-osmosis systems, while preserving valuable mineral salts like calcium, magnesium, potassium and phosphorus, and providing high-quality drinking water.

In areas where excessive hardness is a problem, a high-quality softener/KDF®/carbon block or softener/RO system can be installed for under $2,500. Unless you have unusual water problems, if you are being sold a system that costs more than that, avoid it.

KDF® (COPPER–ZINC) SYSTEMS

This is one of the newest of all the technologies, discovered in 1984 when an inventor named Don Heskett, with over 27 patents to his credit, stuck a brass pen refill into a glass of water colored by a chlorine reagent tablet, and was shocked to see the water's color dissipate. The brass pen refill appeared to be removing the chlorine in some way. Heskett spent the next three years transforming his discovery into a new, patented, water filtration media under the trade name KDF®. KDF® has provided a new way to treat water that shares many benefits of older methods while avoiding their disadvantages.

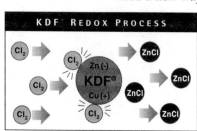

KDF® is a high-purity alloy of copper and zinc that eliminates contaminants from water by utilizing the principle of electrochemical oxidation reduction, known as the redox process. Redox is simply the principle of opposites at work. Some substances are positively charged and are attracted to the negative charge of the zinc, and some substances are negatively charged and are attracted to the positive charge of the copper.

Water, containing dissolved oxygen, minerals, metals, and organic materials, enters a bed of high-purity copper-zinc granules. The copper becomes the cathode and the zinc becomes the anode within each granule. The minute space between each granule becomes an electrolytic cell, with the water and its contaminants acting as the electrolyte. As contaminants and oxygen pass through, some minerals (like heavy metals) adhere to the surface of the granules. In addition, other contaminants react, causing zinc oxides, sulfates, and hydroxides and copper hydroxides, to form in controlled amounts. These compounds are carried into the filtered water and, along with ozone created in the process, provide hostile conditions for algae, fungus, and bacteria growth. The copper and zinc chelates provided by the system are not only harmless in the amounts produced, they are nutrients important to human health.

> **KDF® units are normally combined with activated carbon, which works best at removing what KDF® won't.**

Of further benefit, KDF® works to modify the electron structure of hardness ions, changing them into a very soluble carbonate which does not adhere to surfaces and is easily flushed away, thus interfering with scale deposits. KDF® effectively removes chlorine, chlorinated hydrocarbons, hydrogen sulfide, iron, and heavy metals like lead, arsenic, aluminum, mercury and cadmium. Its only weakness is that it does not do much to remove organic contaminants. That's why KDF® units are normally combined with activated carbon, which works best at removing what KDF® won't.

Because the carbon is installed after the KDF® in the system, the carbon is not called upon to remove chlorine, greatly increasing its useful life. And since KDF® creates a bacteriostatic environment, the carbon's bed does not become a breeding ground for harmful bacteria. Also, carbon block cartridges are available that filter anything over .4 micron in size, making the system an effective filtration system for dangerous biological contaminants like Cryptosporidium and Giardia oocysts.

Independent laboratory tests confirm user experience that KDF® is one of the best tools for improving drinking and bathing water naturally and economically. Compared to carbon-only units, KDF® lasts far longer, doesn't permit bacterial growth, and removes a much wider range of inorganic matter (e.g. heavy metals). Compared to reverse-osmosis, KDF® is less expensive, wastes no water, does not require membrane replacement, works in most water temperatures, and removes chlorine. Compared to ultraviolet lights, KDF® works in turbid water, doesn't require bulb replacement or electricity, and takes out inorganics. Compared to ozone, KDF® lasts longer than 17 seconds for a continued residual bacteriostatic effect in water, does not require electricity, and costs less. Compared to water softeners, KDF® reduces hard scale and helps to condition water without the need for brine tanks, or salt replacement. It is also less expensive to install and operate.

In 1992, the EPA ruled that KDF® qualifies as a mechanical device which filters water and imparts nothing harmful to the water. Later that year, the National Sanitation Foundation tested the media and found it to be in compliance with its Standard 61, which certifies that the media imparts nothing harmful to the filtered water.

Since its introduction, KDF® has found wide application among commercial and industrial users who need clean water for their operations. Breweries and bottling companies have turned to the technology in increasing numbers, as have livestock and poultry producers. In 1992, companies began to manufacture units designed for the home. In the last two years, the home market has grown rapidly, and now KDF® units are available to fit the needs of almost anyone, from inexpensive portable units for apartment dwellers, to whole-house systems. KDF® is effective, affordable, and easy to maintain. Because of its unique combination of benefits, it represents a significant advance in water conditioning devices, and emerges as a major component in the technology of choice for most applications.

NANOFILTRATION   A close cousin to reverse-osmosis membrane technology, nanofiltration (NF) is the newest and one of the fastest-growing segments of the membrane business. In fact, an NF membrane can be visualized as an RO membrane with slightly larger pores.

The term "nano" means one billionth of a meter, and is the theoretical pore size of the membrane. As a reference, a human hair is about 40 microns (millions of a meter), or roughly 100,000 nanometers (nm) in diameter. A typical bacterial cell averages around 1,000 nm, and a helium atom has a diameter of about 0.1 nm. To put nanofiltration in perspective, picture a square foot of membrane as being the size of the Pacific Ocean. At this scale, an RO pore would be roughly the size of a dime, an NF pore would be about the size of a half-dollar. Ultrafiltration (UF) would have a pore size of about an old silver dollar.

Nanofiltration can filter particles as small as a molecular weight (mw) of 300, making it useful for filtering organics, bacteria, parasites and viruses, as well as certain dissolved minerals which account for hardness of water. In fact, it removes almost as much hardness as RO, with a pressure requirement of around half of that required for RO. Both can remove large molecules which create color in the water and contribute to the formation of trihalomethanes.[13]

Both nanofiltration and reverse-osmosis membranes are created by coating a thin layer of a porous polymer onto a backing material. The openings are so small that pressure is required to drive fluid through them. RO filters most inorganic salts from water, and rejects some forms of non-ionic organic compounds, such as fructose molecules (mw = 180), but will pass smaller organics such as ethyl alcohol (mw = 46).

Reverse Osmosis   Nanofiltration   Ultrafiltration

If one square foot of membrane were the size of the Pacific Ocean, a RO pore would be the size of a dime, a NF pore the size of a half dollar and an UF pore the size of an old silver dollar.

As pore size increases, larger and larger molecules will pass through, including more salt. At the point where all salt passes, the membrane is classified as a nanofiltration membrane. Nanofiltration membranes have a pore size between reverse-osmosis and ultrafiltration. Originally known as "high salt passage RO" or "loose RO," nanofiltration rejects organic molecules ranging between 300 to 1,000 mw (molecular weight).[14]

Nanofiltration technology is becoming increasingly important because of its ability to soften water without the need for salt. That's because NF is very effective in rejecting divalent salts such as Ca++ and Mg++ that cause scale and other hardness problems. The end result is water that has been softened without elevating sodium levels or sending a stream of concentrated brine back into the sewer, and eventually, the water someone else will use downstream.

Currently, the largest users of NF technology are municipal drinking water plants and the beverage industry, both of which use the membranes for softening, rather than conventional lime softening. It is also well-established in the dairy industry (where it is used for cheese-whey desalting), and for RO pretreatment, kidney dialysis, and maple sap concentration.

Due to its cost, it is likely to be awhile before nanofiltration units start replacing point-of-use, salt-based water softeners in homes, but prices are dropping fast. Some systems are already beginning to appear.

OZONE   Normal atmospheric oxygen (O_2) exists as a diatomic molecule. This means that two oxygen atoms normally stick together and act as one. Ozone is a triatomic form of oxygen (O_3), that is sometimes called "active" oxygen because, unlike O_2, it is very unstable and highly reactive. Ozone is a gas that is formed when the oxygen in our air is exposed to high intensity ultraviolet light, as our sun creates the ozone layer around the earth. It is also produced by electrical discharge such as lightning or corona discharge.

Ozone is a powerful oxidizer, deodorizer, and disinfectant. It is a safe and highly effective alternative to traditional chemical water treatment systems, and has been used to treat water since 1903. It is a fast-acting disinfectant that functions by rupturing the cellular structure of biological organisms, thus killing them. Air is the oxygen source for generating ozone.

Because ozone is a component of smog, it is sometimes mistaken for smog or air pollution. This is not true. It is the fluorocarbons, hydrocarbons, carbon monoxide and other pollutants in the air that endanger the earth's protective ozone layer. The federal Occupational Safety and Health Administration (OSHA) has set safety standards for ozone concentration in the air we breathe. By their standards, the amount of ozone needed for effective water purification is insignificant.

> Ozone is a powerful oxidizer, deodorizer, and disinfectant. It is a safe and highly effective alternative to traditional chemical water treatment systems, and has been used to treat water since 1903.

Ozone can perform many functions in the treatment of potable water. The origin of most tastes and odors in water supplies is either naturally-occurring organic materials or synthetic organic compounds. Ozone is able to oxidize organic contaminants more completely and more quickly than chlorine. Also, ozone use prevents the formation of chlorine's disinfection by-products like THMs, which are suspected carcinogens.

Ozone is produced by passing dry air, containing oxygen, through a high-frequency electric field. The rate and efficiency of ozone production is affected by many variables, such as air flow, percentage of oxygen in the air feed, dew point of the air feed, air pressure, cooling water temperature, voltage and frequency applied to electrodes, and ozone concentration.

Ozone is very effective as a water conditioner, in that it has the ability to quickly oxidize such nuisance compounds as iron, manganese and sulfur, which cause odor, taste, and staining. These compounds are oxidized to their insoluble forms so they can subsequently be filtered. Ozone is not able to soften water in terms of calcium and magnesium hardness. These elements possess only one oxidation state and therefore cannot be oxidized to an insoluble form. Ozone is also unable to further oxidize nitrates and sulfates, which are soluble at their highest oxidation states and therefore not affected.

Ozone is a much stronger disinfectant and sterilizing agent than chlorine, with a bacteria kill rate 3,000 times that of chlorine, and virus inactivation occurring at an even faster rate. Unlike chlorine, ozone actually breaks down the cellular structure of bacteria, viruses, and cysts in order to deactivate them. The process of chlorine disinfection relies on absorption of the chemical by the organism in order to poison it and disrupt its normal cell function. Chlorine does possess the advantage in that it has a residual action in the water that prevents the re-growth of biological contaminants. Ozone in water has a half-life of just minutes, and it is for this reason that municipal water treatment plants and swimming pools still use chlorine to ensure disinfection.

Ozone is an excellent and proven technology which is widely used in Europe in preference to chlorine for the disinfection of water. In the United States, its widest application is in the disinfection of hot-tub and pool water where it remains a little more expensive, but definitely the treatment of choice. Some home water systems have been designed to use ozone to treat drinking water, but it must be combined with other filtration/purification techniques to make it suitable for the creation of pure drinking water. Ozonated water has a fresh, clean, sweet fragrance resembling pure mountain spring water. It even feels different, like soft, pure rain in the springtime.

> **Ozone is an excellent and proven technology which is widely used in Europe in preference to chlorine for the disinfection of water.**

PORTABLE FILTERS   Most portable filters are made of a durable ceramic material housed in a lightweight case. A reliable unit is excellent at removing bacteria, cysts, algae and sediment from water, but little else. For more information, see *Ceramic Filters* in this chapter.

REVERSE-OSMOSIS   Osmosis is a natural phenomenon whereby the lighter of two substances migrates through a semipermeable membrane toward a heavier solution. This is the process by which our cells absorb nutrients, our lungs transport oxygen to our blood, and plants extract moisture and nutrients from the soil. In reverse-osmosis (RO), the heavier solution is forced under pressure through the membrane, leaving behind those molecules too large to pass through.

RO membranes work on the principle of rejection. The higher the rejection rate, the purer the water produced. The first man-made semipermeable membranes emerged in the 1950s, but were poor performers. Since then, membranes have been refined, and certain types are now capable of very discerning filtration, but they waste a lot of water in doing their job.

RO membranes for residential applications can be divided into two general categories. Thin Film Composite (TFC) membranes have high rejection rates and make very high quality water, but they are easily degraded by chlorine. Cellulose Acetate (CA) or Cellulose Tri-Acetate (CTA) membranes are less expensive and resistant to chlorine, but have a lower rejection rate and therefore don't purify the water as well. They are also easily damaged by bacteria in the water.

A major question about RO systems using CA or CTA membranes recently surfaced through the research of Gene Shaparenko, owner of Aqua Technology Water Stores, a company which manufactures distillers. He discovered that a chemical, known as 1,4 dioxane, is applied to the cellulosic materials of which CA or CTA membranes are made.[15] Its purpose is to etch the small pores in the material through which the water is filtered. 1,4 dioxane is a colorless solvent which mixes easily with water. It is known to be extremely toxic when inhaled or absorbed by skin contact. It is ranked alongside asbestos, benzene, carbon tetrachloride, DDT, formaldehyde, mustard gas, PCBs, TCE, and vinyl chloride on the State of California's Toxic Chemical List.

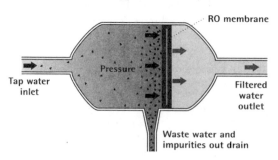

How a Reverse-Osmosis Membrane Works

According to Mr. Shaparenko, no one at the state or federal level of government seems to know that this chemical has been routinely used to manufacture membranes which are used to "purify" water.

Nor does anyone seem to know just how much of the chemical leaches into the finished water product, or how many gallons it takes to effectively purge the chemical, if indeed, it is purged at all.

This is a major concern, because the vast majority of RO systems sold for residential use have employed CTA membranes. Not only are they cheaper than TFC membranes, but resistant to chlorine, and most residential customers are on chlorinated water supplies. As a result, it is possible that many people who have purchased such systems in order to protect themselves have really introduced another deadly toxin into their bodies.

Performance of any type of membrane will depend upon water pressure and temperature. The higher of either, the better the performance of the RO system. Home RO units require normal household water pressure (minimum of 40 psi) in order to operate. At that pressure, you can expect to get around 5 to 7 gallons of water a day from an average unit. The colder the water, the lower the yield. If house pressure is not at least 40 psi, a special pump must be installed to boost the pressure.

RO systems using TFC membranes are extremely efficient at removing 90% to 98% of organic and inorganic chemicals, asbestos fibers, and biological contaminants. They also require no energy to operate except for normal home water pressure. For that reason, they produce water at one of the most inexpensive costs per gallon of any of the technologies. They do, however, waste water, at the rate of 4 to 9 gallons for every gallon of drinking water produced. This problem, coupled with the fact that the waste water concentrates harmful impurities and reintroduces them into the environment, has made the technology illegal in some water-starved localities.

> Reverse-osmosis systems have a penchant for growing bacteria, apparently because the finished water is devoid of chlorine and takes so long to be made.

But there is another, potentially more serious problem with reverse-osmosis water filtration systems. According to a study published in Canada by Canadian virologists,[16] it was discovered that RO systems have a penchant for growing bacteria, apparently because the finished water is devoid of chlorine and takes so long to be made. Because the water sits stagnant at room temperature, sometimes for days, RO systems become an ideal breeding ground for bacteria, which can double their populations every 20 minutes or so. In their study of many types of RO systems, the virologists measured an average of 10,000 bacteria per milliliter of water—20 times the accepted level for city tap water. About a third of the filters produced

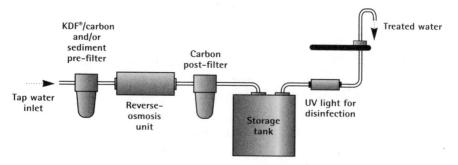

Schematic of an Ideal Reverse–Osmosis System

water with more than 100,000 bacteria per ml, or 200 times more than average tap water. Some bred up to 10 million bacteria per ml!

The study went on to further demonstrate that, because of this contamination (which occurs because bacteria contaminating the end of the dispensing faucet can be "sucked back" into the storage reservoir), people who drink RO water suffer 10 times the number of bouts of gastroenteritis as people who drink tap water. In other words, their expensive filter gives them diarrhea rates like those in Third World countries.

So if you use RO, replace old CA or CTA membranes with TFC membranes, and make sure you replace your pre-filters with a KDF® filter to ensure chlorine does not degrade them. Also, install an ultraviolet light unit between the storage tank and water dispensing faucet.

If you are considering the purchase of an RO system, the best employ a KDF®/granular-activated carbon pre-filter, followed by a thin-film composite reverse-osmosis membrane, followed by an activated carbon post-filter, and then the storage tank. Another carbon filter polishes the water before dispensing, and an ultraviolet light keeps unwanted bacteria from colonizing the system.

RO systems work best when they are working close to capacity. The longer the water stays in contact with the membrane, the more contaminants leak through. Don't order one designed to create more water than you will normally use. Because of the aggressive nature of reverse-osmosis water, minerals should be added back to the water for long-term use. A pinch of ordinary sea salt per gallon will restore mineral balance and improve the taste of the water.

In areas of high hardness, most RO systems work best when preceded by a water softener. By itself, a high quality RO system can be obtained for around $600 to $800, including normal installation. RO systems are the system of choice when water contains high levels of salt that cannot be removed by a KDF®/carbon block system.

ULTRAVIOLET (UV) LIGHT With the invention of mercury arc lamps in 1901, it became possible to disinfect water with UV light. When an organism absorbs light of 254 nanometers, photochemical changes take place which kill the organism. UV is popular because it doesn't require the addition of chemicals to the water, so disinfection by-products aren't formed.

The best UV systems are made of low-pressure, mercury-vapor lamps made of hard quartz glass and using two single-pin, cold-cathode type electrical connections. These types of lamps last approximately 16,000 hours.[17] Quartz sleeves should be used as a jacket because they maintain uniform lamp operating temperatures and transmit more light to the water.

Major factors affecting performance of UV systems are the intensity of the light (called the dose), the contact time, and the clarity of the water to be treated. If the water is murky or turbid, or passes through the system too quickly, the UV light cannot maintain enough contact time with pathogens to kill them, nor can it kill many organisms that are in the water in the form of oocysts. Nor can UV light do anything about toxic chemicals, heavy metals and other pollutants. As a result, UV must always be used as an accessory, and not a primary treatment for drinking water.

9

Whatever you do, if you want to get healthy and stay healthy, don't delay. Make the investment in a quality filtration system and begin to drink large volumes of clean, life-giving water each and every day. Pure, clean water is not a luxury, it is a necessity. A quality water purification system is one of the best investments you can make for the health and well-being of yourself and your family.

SHOPPING TIPS

UNFORTUNATELY, MANY AMERICAN BUSINESSES seem to believe that the only way to market products is through the use of slick, high-pressure tactics. The water conditioning business is no exception. You may have to deal with these aggressive sales techniques when considering the purchase of a water treatment system. Or you may find that you are on your own dealing with salespeople who know little or nothing about the products they are selling, or the real problem with the water contaminants that need to be removed.

Finding a company that prefers to use a true, consumer-oriented educational approach may be difficult, but don't give up. This is a business where every customer's needs and situation can be quite different. Only a company willing to take the time to understand you and your needs, and educate you about potential solutions, is likely to offer long-term satisfaction and value. Avoid firms that use high-pressure sales tactics to sell their product. Competent firms have no need for these tactics. If you are dealing with people who use them, it is best to shop elsewhere.

> **Avoid firms that use high-pressure sales tactics to sell their product. Competent firms have no need for those tactics. If you are dealing with people who use them, it is best to shop elsewhere.**

Don't focus on the company's name or you may regret it later. Some of the biggest companies are the worst. They offer free movie tickets, lamps, clocks, or other paraphernalia to get their salespeople in the door. Many offer free water tests, but the only tests performed are really part of a bigger marketing ploy to sell you a water softener combined with a reverse-osmosis system. Don't be fooled.

Before you let representatives from any company in your door, ask them what they will test for and the cost of the water test, if any. Many legitimate companies charge for their tests because comprehensive tests are expensive. Some may offer to credit the cost of the test to the purchase of a system if you buy from them. That is fine. But if the only test a company offers is for hardness or the presence of chlorine under the name of a water test, expect a sales job, not an education. They are probably using deceptive sales tactics.

Once a company representative is in your door, observe his/her attitude. While it is natural to expect anybody to want to make the sale, if the sale comes before your needs, it is not a good idea to do business with that company.

After some friendly chit-chat and the demonstration (test), the time will come for the salesperson to ask you to buy. That is normal and understandable. But beware of deceptive closing tactics. For example, you may be told that their brand-name product has the best warranty in the industry. You may even be told that if you can find other equipment with the same warranty for less money, the company will give your money back and let you keep the equipment! Sounds good doesn't it? That's precisely why they do it. They want to motivate people to buy without shopping the competition.

If you hear this from a company, don't buy until you do check the competition. You may find that the "best warranty in the industry" does not really give you the protection you expect. And, of course, other companies' warranties are worded differently; no two companies word their warranties exactly the same. So you won't be able to get your money back and keep the equipment. Such a technique is just a form of deceptive marketing.

Other closing techniques may include a free supply of soap products, free installation, free sales tax, or some other bonus. That's fine, too. But if the only way you get the bonus is to buy right then without thinking about it or shopping the competition, it's deceptive marketing. Remember the old adage, if it sounds too good to be true, it probably is. Deal only with companies that tell you the truth. There is no free lunch. You may count on the fact that you are paying for all the freebies somewhere.

Also beware of financing plans. Companies know that people are more concerned with the down payment and monthly payment than the overall price. Read the fine print! You have no bargain if you must pay 18% interest for five years on a loan to purchase the equipment. The payment may look okay, but the deal will be outrageously expensive in terms of interest paid over the duration of the loan. Again, you may be offered a free downpayment or no payments for the first three months as an inducement to buy. Remember, you will be

BUYER BEWARE!

1 Watch for sales gimmicks and promotions
2 Compare the competition's prices, systems, warranties
3 National brands not necessarily best
4 Beware of uninformed salespeople
5 Beware of financing plans

paying for it somewhere down the line, with interest—nothing is free!

Finally, carefully scrutinize products offered by multi-level or pyramid marketing outfits. People who work for these companies are offered visions of wealth and an abundance of sales training, but product training is scarce and largely up to the individual. That means many of the people who sell these products have only enough information to be dangerous. Since the commission structure with these firms is, by necessity, very high, most of the time these products deliver more hype than performance for the dollar.

FINDING VALUE

So what do you look for if you want to find the best value? First, realize that value is less about how much you spend for the system than about how much you spend *per gallon* for finished water of a quality that meets your needs. Once you are assured that the water produced will meet your needs, divide the price of the unit installed, plus the cost of maintenance over the system's projected life span, by the number of gallons it is designed to treat.

Find a company whose salespeople are knowledgeable and willing to take as much time as necessary to educate you. If you have read this book you can ask intelligent questions; and, if the salespeople (company) are knowledgeable, you will know it by their answers. If they are there to help rather than just make a sale, you will know it by their attitude. If their products are truly good values and they know it, they won't need to make you offers of free this or that in order to induce you to buy, and they won't need to pressure you to buy before you check out the competition.

If you need financing, most banks and finance companies love to make home improvement loans at competitive interest rates. As with the equipment itself, shop around. A few phone calls will suffice to hook you up with a lender that won't line its pockets at your expense. Reputable companies will be able to steer you to hometown lenders who will be more than happy to have your business.

WHAT TO LOOK FOR

You are looking for a system that not only removes, but will not breed, harmful bacteria or other disease-causing organisms. The system should remove sediment, turbidity, chlorine and its associated DBPs (disinfection by-products like trihalomethanes and chloramines), nitrates, chemical pollutants, heavy metals, excess iron and manganese, hydrogen sulfide (if present), odors, and bad tastes. The system should consume very little energy, and remove contaminants at a very low cost per gallon, with little maintenance. Finally, you want the system to be available for a reasonable initial investment.

At the time of this writing, the best overall choice for most people appears to be a dual-stage system which uses a KDF®/granular activated carbon cartridge to remove chlorine, its associated DBPs, VOCs and other harmful organic and inorganic chemicals (i.e., PCBs and PBBs), and heavy metals, combined with a second stage carbon block cartridge (.5 micron or smaller *absolute*) that will block the cysts of organisms like Giardia and Cryptosporidium. The KDF® pre-filter will aid in keeping the carbon media bacteriostatic, and help it to last far longer. If hardness is a problem, invest in a quality water softener.

Systems which treat all the water in the house using ozone in combination with a water softener and KDF®/carbon block filters, in addition to UV disinfection before dispensing, are the ultimate. You want the system to use standardized cartridges, and it is best if it employs a meter or some other reliable method to notify you when it is time to change the cartridges.

The best of these systems produce ideal water in which harmful contaminants are virtually eliminated, while the healthful mineral salts are retained. The water tastes better and is better for you than RO water.

As of this writing, the best systems cost around $600 to $1,000 installed, but many perfectly adequate systems are available for $400 to $600. That's assuming you don't need a water softener or don't need purified water at every tap in your house. Whole-house systems can cost $3,000 to $5,000 installed, or more.

If the water is unduly salty or has some other unusual problem, your second choice is an RO system which uses a thin-film composite (TFC) membrane, a KDF®/carbon prefilter, and a UV light before the tap. A quality RO system will produce highly purified water, but will waste a lot of water in the process (10,000 to 15,000 gallons per year or more). The water will not be as healthful as that produced by a KDF®/carbon system, and the system will cost more to install and maintain. Expect to pay between $650 to $850 for a quality system, although many large name-brand companies sell lesser quality systems for much more.

SUMMARY

It is disturbing to realize that a major percentage of the nation's drinking water supply doesn't meet safety standards set in 1962, much less the current, stricter standards, which most researchers agree are far too lenient in some areas. It is also disturbing to realize that most standards are routinely violated because little or nothing is done to enforce them.

Current research demonstrates that chlorine, almost universally added to water supplies to kill bacteria, frequently forms potent cancer-causing chemicals as a by-product. EPA reports to Congress estimate that all U.S. public water supplies contain chloroform, and that many supplies contain an abundance of chemical contaminants.

Until recently, the public has been more or less uninformed about these issues, a situation which some regulators would like to maintain. But that is changing. The public has begun to wake up as entire cities have experienced outbreaks of infectious disease, and thousands of people have become ill from drinking toxic water. Bad water is making big news these days.

Responding to the reports about these problems, and perceiving a major money-making opportunity, dozens of companies have recently brought out a large number of point-of-use drinking water purification products. Many of these units are sold as "purifiers," although there are significant differences as to what these various units will remove. Many are inappropriate for their intended use; some don't seem to be effective at removing much, if anything. But this doesn't stop the advertising claims, some of which seem to be limited only by the imagination of the advertising people hired by the company making the claims.

Recently, the EPA has specifically defined "purify" to mean "render safe from harmful bacteria," but most of the products, in fact, do not protect against harmful bacteria and cysts. That's why many companies have changed their product's description to something like "water processor," "naturalizer," or "water treatment system." Even more confusing, some companies making these non-purifying products use a variant of the word "pure" incorporated into brand or company names, even registering these names with the EPA. Although EPA regulations forbid this type of misleading advertising, the EPA doesn't seem to be interested in enforcing its own rules in this regard, and enforcement in general has, at best, been inconsistent. As always, it falls to the consumer to become educated about the problem and take whatever measures are necessary to safeguard the health of themselves and their families.

Unit	Advantages
DISTILLATION SYSTEM	Removes all inorganic salts and destroys bacterial and viral contaminants. Beneficial minerals can be added back to drinking water.
ACTIVATED CHARCOAL FILTER	Inexpensive. Removes foul tastes and odors, some carcinogens (cancer-causing agents) and hazardous chemicals (insecticides, chlorine, chloroform, iodine, formaldehyde)
BACTERIOSTATIC CHARCOAL FILTER *Must be EPA registered*	Suppresses growth of bacteria within charcoal bed.
CARBON BLOCK FILTER *.5 micron absolute or less*	Removes foul tastes and odors, most carcinogens and hazardous chemicals, bacteria, cysts, asbestos, and other fine particles. Catalytic carbon will remove hydrogen sulfide as well. Prevents water from "channeling" ensuring proper contact time.
CARBON BLOCK FILTER WITH DEIONIZING RESIN *.5 micron absolute or less*	All of the advantages of carbon block filters above. Will remove most heavy metals as well.
KDF® FILTER	On of the most effective means of removing chlorine, chloramines, and other DBPs. Removes heavy metals. Does not remove healthful minerals. Releases beneficial zinc and copper chelates into the water. Renders water bacteriostatic.
KDF®/CARBON BLOCK SYSTEM *.5 micron absolute or less*	Combines all the advantages of carbon block and KDF®. Produces water as free of harmful contaminants as RO while retaining healthful minerals. Less expensive than RO. Does not waste water. Inexpensive to maintain. Does not foul as easily as RO in hard water.
REVERSE-OSMOSIS SYSTEM	Removes nitrates, asbestos, fluorides, sodium and other dissolved salts. Removes heavy metals, chlorine, and DBPs. Can produce high quality drinking water. Minerals can be added back to product water.
ULTRAVIOLET PURIFIER (UV)	Effective in destroying bacteria when operating at full efficiency on clean water. High flow capacity. Relatively low power consumption.
OZONIZATION SYSTEM	Effective in destroying bacteria, some viruses, and many substances that impart foul tastes and odors to water. Oxidizes iron, manganese and hydrogen sulfide, reducing them to insoluble forms so they can be filtered. Oxidizes many volatile organic chemicals.
BOTTLED WATER	May be better than tap water. Some mineral waters have high levels of healthful minerals. Convenient to purchase when away from home.

Requires electrical connection. Some units may concentrate volatile chemical contaminants. Uses excessive amounts of energy. Creates heat. Output relatively low.

Can support runaway multiplication of bacteria. May remove symptoms or chemical contaminants without removing the contaminants. Requires sufficient holding time to be effective. Does not remove bacteria, cysts, heavy metals, nitrates, asbestos, and other fine particles.

Same disadvantages of ordinary carbon filters regarding chemicals, cysts, heavy metals, nitrates, asbestos, and other fine particles. No way of knowing when effectiveness against bacteria is depleted. May add silver or other pesticides to the water.

Does not remove heavy metals, nitrates, and the disinfection by-products (DBPs) of chlorine such as trihalomethanes and chloramines. Does not remove hydrogen sulfide. Easily plugged if pre-treatment is inadequate.

Relatively expensive. Does not remove all DBPs. Can become rapidly depleted if water is high in nitrates and chlorine. Easily plugged if pretreatment is inadequate.

Relatively expensive. Does not remove bacteria, viruses, cysts, asbestos, and other fine particles. Does not remove all volatile organic chemicals, pesticides or herbicides.

Does not remove high levels of sodium chloride. If water has high levels of sodium, or other unusual problems, RO with TFC membrane, KDF® pre-treatment, and UV disinfection becomes the treatment of choice.

Relatively low output. Wastes large amounts of water. Does not work well if water is cold and pressure is not at least 40 psi or more. Units with cellulose acetate (CA) or cellulose triacetate (CTA) membranes may leach dangerous 1,4 dioxane into the water. Units with thin-film composite membranes are easily damaged by chlorine so adequate pre-treatment with carbon or KDF® required. Runaway growth of bacteria in storage tank and carbon post-filters necessitates the use of UV light ahead of tap.

Requires electrical connection. Does not remove chemicals, tastes and odors, asbestos, nor any other non-biological contaminants. Efficiency and effectiveness can drop off without warning. Water must be clear with very low turbidity to be effective.

Relatively expensive. Requires electrical connection. Does not remove nitrates, sulfates, or dangerous heavy metals.

By far the most expensive alternative over time. Much bottled water is simply tap water. Often mislabeled. May contain high levels of bacteria because purified, non-chlorinated water is an ideal breeding ground for bacteria. May contain harmful chemicals, either leached from containers or inadvertently introduced at source. Heavy to transport.

❏ System removes harmful bacteria, viruses, and similar organisms as well as the oocysts of Cryptosporidium and Giardia. (Meets NSF Standard 53)

❏ System will not create an environment conducive to the breeding of biological contaminants.

❏ System does not employ reverse-osmosis or carbon without KDF®, ozone, or ultraviolet light protection.

❏ System removes odors, bad tastes, and filterable sediment without channeling.

❏ System effectively removes chlorine and DBPs (disinfection by-products) like trihalomethanes and chloramines.

❏ System effectively removes organic chemical pollutants.

❏ System effectively removes all harmful heavy metals like lead, cadmium, mercury, and aluminum.

❏ System effectively removes excess iron, manganese, and hydrogen sulfide.

❏ System does not remove healthful minerals like calcium, magnesium, phosphorus, and potassium.

❏ System does not consume much energy.

❏ System does not waste excessive amounts of water.

❏ System does not require expensive maintenance, has a system to notify you when maintenance is required, and is easily maintained by the user.

❏ System has a reasonable initial investment.

RESOURCES

For more information on water pollution, water resources and action guides:

American Ground Water Trust
800-423-7748

American Planning Association (APA)
312-431-9100
Publishes *Protecting Nontidal Wetlands* (David G. Burke et al.,1989), and bibliography
Groundwater Quality: Trends Toward Regional Management (1988).

American Water Resources Association (AWRA)
301-493-8600
Publishes *Aquatic Organisms as Indicators of Environmental Pollution* (Joan A. Browder, ed.,
1988), *Redefining National Water Policy: New Roles and Directions* (Stephen M. Born, ed., 1989),
Water Management in the 21st Century (A. Ivan Johnson and Warren Viessman, Jr., eds., 1989)
and *Water: Laws and Management* (Frederick E. David, ed., 1989) also publishes a series of
Regional Aquifer System Analysis (RASA) Program Studies with the U.S. Geological Survey.

American Water Works Association
303-794-7711

Center for Science and Technology for Development (CSTD)
212-963-8435

Chesapeake Bay Foundation (CBF)
410-268-8816
Publishes *Your Boat and the Bay, Septic Systems and the Bay* and *Water Conservation:
Wasted Water Means Wastewater.*

Clean Water Action
202-457-1286

Concern, Inc.
202-328-8160
Publishes *Groundwater: A Community Action Guide,* and
Drinking Water: A Community Action Guide.

Conscious Living Systems, Inc.
505-351-2927
P.O. Box 3690, Santa Fe, NM 87502
Information on effective and environmentally safe water conditioning systems.

Conservation Foundation
202-429-5660
Publishes *News from the Front* (bimonthly)

Conservation Fund, The
703-525-6300
Sponsors the Spring and Groundwater Resources Institute.

Environmental Defense Fund (EDF)
212-505-2100
257 Park Avenue South, New York, NY 10010
Publishes *Polluted Coastal Waters: The Role of Acid Rain* (D. Fisher et al., 1988).

Environmental and Energy Study Institute (EESI)
202-628-1400
Publishes *Farm Policies to Protect Groundwater: A Midwestern Perspective* (1990), and *Groundwater Protection and the 1990 Farm Bill* (1990).

Environmental Hazards Management Institute (EHMI)
603-868-1496; 603-868-1547
Publishes a useful "water sense wheel."

Environmental Law Institute (ELI)
202-328-5150
Publishes *Clean Water Deskbook* (rev. ed., 1991).

Environmental Protection Agency (EPA)
Public Information Center 202-260-7751 or 202-260-2080
Operates Safe Drinking Water Hotline 202-382-5533 (or toll-free outside DC: 800-428-4791).

Freshwater Foundation
612-471-9773
Publishes the following: *Nitrates and Groundwater: A Public Health Concern; Waste is a Water Problem: What You Can Do About Solid Waste; Water Filters: Their Effect on Water Quality; Hazardous Waste in Our Home–and in Our Water; Understanding Your Septic System; Freshwater Journal; Facets of Freshwater Newsletter; U. S. Water News; Watershed Management: A Community Commitment; Groundwater: Understanding Our Hidden Resource.*

Friends of the Earth (FOE)
202-783-7400
Publishes *Bottled Water: Sparkling Hype at a Premium Price* (Sandra Marquardt, 1989); and *Groundwater Newsletter.*

Friends of the River (FOR)
415-771-0400
Conducts Water Policy Reform campaign and Hydromania and Watershed Protection campaign.

Greenpeace
202-462-1177
Greenpeace Action runs the Ocean Ecology Campaign, the Toxics Campaign which publishes *We All Live Downstream: The Mississippi River and the National Toxics Crises* (1989); *Water for Life: The Tour of the Great Lakes* (1989), and *Mortality and Toxics Along the Mississippi River* (1989). Greenpeace also publishes a variety of action fact sheets and videocassettes.

Human Ecology Action League (HEAL)
404-248-1898
Publishes *Water Testing* and *Water Treatment.*

Inform, Inc.
212-361-2400
Publishes *Winning with Water: Soil Moisture Monitoring for Efficient Irrigation* (Gail Richardson and Peter Mueller-Beilschmidt, 1988).

International Center for Arid and Semiarid Land Studies (ICASALS)
806-742-2218
Publishes *Making Rain in America: A History.*

International Council For The Exploration of the Sea
or Counseil International pour l'Exploration de la Mer (ICES or CEIM)
Copenhagen K, Denmark: 33 15 42 25

International Transboundary Resources Center
505-277-6424

Izaak Walton League of America (IWLA)
703-715-6643

National Coalition Against Misuse of Pesticides (NCAMP)
202-543-5450
Publishes *The Great American Water Debate.*

National Environmental Health Organization (NEHA)
303-756-9090
Publishes *Groundwater Quality Protection; Groundwater Contamination:Sources, Control and Preventive Measures; On Site Wastewater Disposal; Septic Tank System Effects on Groundwater Quality; and Groundwater Pollution Control and Drinking Water Health Advisories–Pesticides.*

National Parks and Conservation Association (NPCA)
202-223-6722

National Water Resources Association (NWRA)
703-524-1544

National Wildlife Federation (NWF)
202-797-6800 or 800-245-5484
Publishes *An Environmental Agenda for Clean Water: Prevent, Protect and Enforce.*

National Resources and Energy Division (NRED)
United Nations
212-963-6205

Natural Resources Defense Council (NRDC)
212-727-2700
Publishes *Ebb Tide for Pollution: Actions for Cleaning Up Coastal Waters* (1989); and *Poison Runoff: A Guide to State and Local Control of Nonpoint Source Water Pollution* (Paul Thompson et al., 1989).

North American Lake Management Society (NALMS)
202-466-8550
Publishes the bimonthly magazine *Lake Line,* the bimonthly journal *Lake and Reservoir Management,* and other books, reports and videos.

Northwest Coalition for Alternatives to Pesticides (NCAP)
503-344-5044
Publishes *Every Drop Matters: A Guide to Preventing Groundwater Contamination* (Neva Hassanein and Ivy Cotler, 1989).

Population–Environment Balance
202-955-5700
Publishes *Water Availability and Population Growth.*

Renew America
202-232-2252
Publishes focus papers *Drinking Water* (1989) and *Surface Water Protection* (1988).

Resources for the Future (RFF)
202-328-5000
Publishes *Markets for Federal Water: Subsidies, Property Rights, and The Bureau of Reclamation* (Richard W. Wahl, 1989).

Rocky Mountain Institute (RMI)
970-927-3851
Publishes *Water Efficiency for Your Home: Products and Advice Which Save Water, Energy, and Money* (John C. Woodwell).

Sierra Club
415-776-2211 or 202-547-1141; legislative hotline: 202-547-5550
Publishes the *Hazardous Materials/Water Resources Newsletter* and *Water Policy*.

Soil and Water Conservation Society of America (SWCS)
515-289-2331 or 1-800-THE-SOIL (843-7645)
Publishes a bi-monthly multidisciplinary *Journal of Soil and Water Conservation (JSWC)*; the bi-monthly *Conservogram;* a middle-school-age computer program *Farm and Food Bytes: Soil and Water Conservation;* also publishes books, cartoon-style booklets and guides oriented toward school-age children.

Trout Unlimited
703-522-0200
Publishes *Trout* (quarterly magazine)

United Nations Environment Programme (UNEP)
212-963-8094
Publishes *International Conventions on the Prevention of Marine Pollution: Coastal Strategies* (M. Nauke, 1991), and *Assessment of Freshwater Quality* (Global Environment Monitoring System [GEMS] and World Health Organization [WHO], 1988).

U.S. Geological Survey
703-435-9269
423 National Center, Reston, VA 22092-0001
Provides federal water data.

U.S. Public Interest Research Group (PIRG)
202-546-9707
Publishes *Permit to Pollute: Violations of the Clean Water Act by the Nation's Largest Facilities* (1991).

Water Environment Federation (WEF)
703-684-2400 or 800-666-0206
601 Wythe Street, Alexandria, VA 22314
An international professional membership organization of water-quality experts, organized into a federation of 70 member associations around the world, its aim being the preservation and enhancement of water quality worldwide. Sponsors an annual conference on water quality and pollution control technology and issues; provides administrative support for *Water Quality 2000.* For the general public, it publishes a Water Environment Curriculum with videos and student and teacher guides for grades 5-9, including *Surface Water Unit; The Groundwater Video Adventure; The Wastewater Treatment Video;* and *Saving Water-The Conservation Unit;* the video *Careers in Water Quality* for high school students; and brochures such as *Let's Save the*

Environment; Clean Water: A Bargain At Any Cost; Nature's Way: How Wastewater Treatment Works for You; Groundwater: Why You Should Care; Nature Recycles Water...We Can Too; Clean Water For Today: What is Wastewater Treatment? and *Stop Water Pollution!* It also publishes numerous technical works and materials for wastewater professionals.

World Association of Soil and Water Conservation (WASWC)
605-627-9309 (fax)

World Bank
202-473-5787
Publishes *Environmental Health Components for Water Supply, Sanitation, and Urban Projects; Wastewater Management for Coastal Cities: The Ocean Disposal Option, Water Pollution Control: Guidelines for Project Planning and Financing;* and *Wastewater Irrigation in Developing Countries: Health Effects and Technical Solutions.*

World Resources Institute (WRI)
202-638-6300

Zero Population Growth (ZPG)
202-332-2200
Publishes *Water Wars* and *In Troubled Waters.*

OTHER REFERENCES

Water Resources Planning, Andrew A. Dzuirk. Rowman and Littlefield, 1990.

The Water Encyclopedia, 2nd. ed. Frits van der Leeden et al. Lewis, 1990.

Overtapped Oasis: Reform or Revolution for Western Water, Marc Reisner and Sarah Bates, Island Press, 1990.

Water and the Future of the Southwest, Zachery A. Smith, ed. University of Arizona Press, 1989.

Urban Surface Water Management, Stuart G. Walesh. Wiley, 1989.

Troubled Waters: New Policies for Managing Water in the American West,
Mohamed T. El-Ashry and Diana C. Gibbons. Cambridge University Press, 1988.

Water Law, 2nd ed. William Goldfarb. Lewis, 1988.

Practical Handbook of Ground-Water Monitoring, David M. Nielsen, ed. Lewis, 1991.

Surveillance of Drinking Water Quality in Rural Areas, Barry Lloyd and Richard Helmer. Wiley, 1991.

Drinking Water Quality: Standards and Controls, John De Zuane. VanNostrand Reinhold, 1990.

Global Freshwater Quality: A First Assessment, Michel Meybec et al., eds. Blackwell Reference (Oxford, U. K.), 1990.

NOTES

1 Ehrlich, as cited in *United Nations Statistical Commission and Economic Commission for Europe. The Environment in Europe and North America: Annotated Statistics 1992* (United Nations, New York, 1992). Table II-2.4.6 p. 215.

2 F. Batmanghelidj, M.D., *Your Body's Many Cries for Water,* examines the implications of research exploring the link between disease and chronic dehydration. Available from APMA (800-230-APMA) and bookstores.

3 Reilly, W. K., *Aiming Before We Shoot: The Quiet Revolution in Environmental Policy* (Address delivered September 26, 1990).

4 Natural Resources Defense Council, *Think Before You Drink: The Failure of the Nation's Drinking Water System to Protect Public Health,* (Sept. 1993).

5 Morris, R. D. et al., "Chlorination, Chlorination By-Products, and Cancer: A Meta-Analysis," *American Journal of Public Health,* Vol. 82, no. 7, pp. 955-963 (1992).

6 1975 report from the Water Quality Association to the United States Bureau of Labor Statistics conducted in 1974 by M. L. Birenbaum.

7 Karl H. Schutte, M. Sc., Ph. D., *The Biology of Trace Elements,* J. B. Lippincott and Company, Philadelphia and *Metabolic Aspects of Health,* by John A. Myers, M.D., F.R.S.H., and Karl Schutte, M. Sc., Ph. D., Discovery Press, 1978.

8 G. Morgan Powell, "Magnetic Proof Needed," *Water Technology,* Vol. 18, no. 3, March 1995, pp. 10-12.

9 In a June 1989 conference sponsored by the University of Missouri.

10 Ijsselmuiden, C. B. et al, "Cancer of the Pancreas and Drinking Water: A Population-Based Case-Control Study in Washington County, Maryland," *American Journal of Epidemiology,* Vol. 136, no. 7 pp. 836-842, (1992).

11 Bove, F. J., et al, "Public Drinking Water Contamination and Birthweight, Fetal Deaths, and Birth Defects" U.S.Public Heath Service and N.J. Dept. of Health (1992).

12 As reported in *American Laboratory Magazine,* May 1989.

13 David H. Paul, "Nanofiltration Offers a Membrane Option, It Can Be Less Expensive Than Using RO," *Water Technology Magazine*: Vol. 18, no. 2, February 1995, pp. 38-42.

14 Lee Comb, "How to Soften Water Without Salt, Nanofiltration Sends Hardness Down the Drain," *Water Technology Magazine:* Vol. 18, no. 2, February 1995, pp. 30-36.

15 Gene Shaparenko, *The South Valley Times,* "Lifelines", April 1990.

16 Pierre Payment, Eduardo Franco, Lesley Rishardson, and Jack Siemiatyci, "Gastrointestinal Health Effects Associated with the Consumption of Drinking Water Produced by Point-of-use Domestic Reverse-osmosis Filtration Units," *Applied and Environmental Microbiology,* Vol. 57, no. 4, April 1991, pp. 943-948.

17 Farrukh Saeed Malik, "In Developing World, UV's Future Is Bright, Its Economy and Ease of Maintenance Shine" *Water Technology Magazine:* Vol.18, no. 3, March 1995, pp. 39-40.

INDEX

A

Aeration 20, 50, 53
Agriculture 11
Algae 69
Alkalinity 34, 40
Aluminum *see Contaminants*
Alzheimer's 20
Aquifers 1, 3-4, 8, 10
Arsenic *see Contaminants*
Artesian water 3, 56
Arthritis 20, 23, 36
Asbestos *see Contaminants*
Asthma 23-24
Atherosclerosis 45

B

Backwashing 37, 62
Bacteria 39, 43-44, 57, 59-61, 65, 67, 69, 71
    E. Coli 14
Bacteriostatic 38, 60-61, 65-66, 80
Benzene *see Contaminants*
Biological Contaminants 7, 41
    Cryptosporidium 14, 31, 43, 59, 61, 65
    cysts 43, 69
    Giardia 31, 43, 59, 65
Biological organisms 43
Bladder cancer 46
Bladder stones 36
Bloodworms *see Parasites*
Bottled water 55-57, 80
    expense 57
Bottled Water Association 56

C

Cadmium *see Contaminants*
Calcium 33-34, 36, 52, 68, 82
Cancer 20, 27, 42, 47, 49, 79
Carbon filters *see Filters*
Carbon tetrachloride *see Contaminants*
Carcinogens 27, 46, 56, 68
Cartridges *see Filters*
Catalytic/adsorptive filters *see Filters*
Catalytic oxidizing filters *see Filters; iron*
Cellulose Acetate *see Membranes; (CA)*
Cellulose Tri-Acetate *see Membranes; (CTA)*
Centers for Disease Control 27, 43
Ceramic filters *see Filters*
Charcoal filters *see Filters*
Chloramines 48, 61
Chlorination 20, 27, 31, 37, 40-41, 44-47, 50, 69
Chlorine *see Contaminants*
Chloroform *see DBPs; trihalomethanes*
Coagulation 19-20
Contaminants
    aluminum 11, 20, 31, 48, 65
    arsenic 31, 41, 57, 65
    asbestos 58, 61, 70, 81
    benzene 56, 70
    cadmium 7, 36, 40, 65
    carbon tetrachloride 11, 70
    chlorine 11, 27, 31, 40, 44-48, 57,
        60-61, 64-65, 76, 79
    copper 34, 40, 57
    cyanide 7
    fecal pollution 44
    formaldehyde 70, 80

gasoline 10
heavy metals 7, 9, 61, 64
hydrazine 59
hydrogen sulfide 31, 50, 58, 61, 65
iron 31, 38-40, 50, 57, 65
lead 7, 26, 40, 50, 52, 57, 65
manganese 31, 40, 50, 57
mercury 7, 57, 65
minerals 7, 41, 62
nitrate nitrogen 52-53
nitrates 10, 31, 35, 52, 57, 61, 69
phosphates 10, 35
radon 7, 31, 53
silver nitrate 59
sulfates 41, 57, 69
uranium 7, 53
Contamination 4, 7, 15, 26-27, 51-52, 55, 57
Copper *see Contaminants*
Copper chelates 65, 80
Corrosion control 19
Cryptosporidium *see Biological Contaminants*
Cyanide *see Contaminants*
Cyanosis 53
Cysts *see Biological Contaminants*

D

DBPs 40, 46, 48, 60, 64, 68
    chloramines 60
    trihalomethanes 27, 40-41, 46-48, 58,
        60-61, 67
        chloroform 46, 58, 61
        trichloroethylene 46
DDT 70
Dehydration 23-26
Deionization 42, 53
Disinfection 19, 44, 47
Disinfection by-products *see DBPs*
Distillation 43-44, 48, 50, 53, 56, 61-62, 80
    fractional 62
Drinking Water Standards 14, 56
Dust mites 57

E

Ecosystems 8
Edema 25
EPA 3, 9-10, 13-15, 17, 26-27, 40, 42, 46, 51,
    55, 57, 66, 79

F

Fecal pollution *see Contaminants*
Fertilizers 8, 11
Filters 82
    activated carbon 40, 42, 48, 50, 53, 58-60, 65
    activated charcoal 80
    carbon block 43, 59, 61, 64-65, 80
    cartridges 60
    catalytic/adsorptive 58
    ceramic 43, 58, 61
    expense 64, 76, 78
    catalytic oxidizing 39
    KDF® 38-39, 42, 44, 47-48, 53, 60-61,
        63-66, 78, 80
    portable 69
Filtration 19, 43-44, 47, 52, 61-62, 64-65, 69-71
Flocculation 19-20
Fluoridation 20, 27, 48-49

A Word About the Author

LONO IS A KAHUNA from the Big Island of Hawaii. His interest in natural medicine began with conventional medical training at the University of Colorado. It was there he was both fascinated by and trained in the basic sciences of the physical world, taking many courses in various branches of biology, chemistry, and physics. It was also there he learned about western society's approach to healing and chose to avoid making western allopathic medicine a lifetime career.

That decision led Lono to pursue a path which eventually resulted in his becoming a kahuna. As a kahuna, he has dedicated his life to the recovery of old world wisdom and re-establishment of life-giving social structures based on family.

Lono has spent over 30 years studying and practicing the art and science of enlightenment as it applies to economics, natural medicine, meditation, prayer, breathing, and other conscious living skills. After operating a healing and retreat center on the Big Island of Hawaii, he moved with his beloved wife Ka Ike Lani, to northern New Mexico. There they are busy developing and operating the Eagle Medicine Healing and Retreat Center near Santa Fe.

To Lono, an abundant supply of life-giving water is an essential element of physical well-being. He has written this guide so that all his brothers and sisters may know of water's importance to health, and how to ensure that they are drinking something healthy, instead of something poisoned.

Specializing in alternative health-related topics

Additional copies of **Don't Drink the Water** are available through Kali Press.

ISBN 0-9628882-4-9

$14.95*

The Authoritative Tea Tree Oil Reference Books

Author/researcher Cynthia Olsen presents the most comprehensive books on this ancient remedy. The **Manual** is for those who want to get acquainted with the oil. The **Guide** contains information about production, quality and research results. The **Handbook** describes 101 ways to use tea tree oil (*Melaleuca alternifolia*) from head to toe—a must for users of this "first aid kit in a bottle."

Manual ISBN 0-9628882-0-6 $3.95*
Guide ISBN 0-9628882-1-4 $6.95*
Handbook ISBN 0-9628882-2-2 $2.95*
Handbook (in Spanish) ISBN 0-9628882-3-0 $3.50*

Watch for upcoming titles from Kali Press including a guide to the herbal remedy **Essiac**.

Little Kali Press publishes children's books. Our titles include: **A Wasco Indian Story: The Spirit of Gray Elk,** and **Messenger of Love, an Angel Story.**

*Shipping and handling is extra. Please call for prices.

Bookstores contact Kali Press for ordering information:

Kali Press
P.O. Box 2169
Pagosa Springs
Colorado 81147
Telephone 970-264-5200 / Fax 970-264-5202